RANDOM VIBRATION

IN PERSPECTIVE

RANDOM VIBRATION
IN PERSPECTIVE

Wayne Tustin and Robert Mercado

TUSTIN INSTITUTE OF TECHNOLOGY
SANTA BARBARA · CALIFORNIA

Acknowledgment.

Books require much effort by many people over several years. Specific technical contributions are acknowledged in many sections. TIT's office staff not only operated our normal business (private engineering school) but also many times revised the book as it developed. The authors greatly appreciate the help of, most particularly, Barbara Altman (TIT's office manager) and her predecessor, Jenean Thompson. Barbara and Jenean were assisted by Carolyn and Janet Mercer. The authors also thank their wives, Shirley Tustin and Linda Mercado, whose patience made the writing possible and whose proofreading caught many errors. Tom Leja consulted on the book design and layout. Joe Stewart and his team (particularly Scott Abbott, Robin Ryan and Tom Franotovich) at Santa Barbara Graphics who did the typesetting, layout and paste-up, and camera work, respectively; Fred Gerhart negotiated and supervised actual production.

© 1984 by Wayne Tustin

First edition, 1984
Library of Congress Catalog No. 84-080801
ISBN 0-918247-00-4

Tustin Institute of Technology, Inc.
22 East Los Olivos Street
Santa Barbara, California 93105 USA
Printed in the United States of America.

Other titles by Wayne Tustin:
Environmental Vibration and Shock Testing, Measurement,
    Analysis and Calibration, 1962, revised 1979.
Vibration and Shock Test Fixture Design, by Klee, Kimball
    and Tustin, 1971.

Preface

How this book came to be written

"Random Vibration in Perspective" began in 1962 as a series of lessons by mail. They dealt with vibration measurement, analysis, calibration and testing. Author and TIT founder Wayne Tustin managed to stay about a week ahead of his fastest students. In 1963 Holloman Air Force Base, New Mexico had Tustin come to teach "live." The lessons had to be repackaged into an informal text. With much rewriting, "cutting and pasting" and adding new material, the text was used by thousands of TIT students 1963-1983.

But a complete rewrite was needed. (There is a limit to how many times one can patch a garment.)

One reason for delaying the new text was the "D" revision to Military Standard 810. Its official date is 19 July 1983. 810D demands vastly more random vibration measurement, analysis and testing. Sinusoidal vibration testing barely survives.

Authors Mercado and Tustin would like to dedicate this text to the many recognized and unrecognized individuals, in Government and in private industry, who contributed to 810D.

The environmental engineering specialist (EES)

What is an "environmental engineering specialist?" 810D defines him as someone who must "choose and alter the test procedures to suit a particular combination or sequence of environmental conditions for a specific equipment application." He is the environmental engineering "expert." He works with the technical aspects of natural and induced environments. He measures and analyzes environments and formulates environmental test criteria.

In the past, test requirements and design requirements have often been totally different. Usually the test requirements won out because they had to be passed to satisfy a contract. Thus equipment was usually designed to pass a test, regardless of what was needed in the field. Under 810D, "tailoring" tests to match the service environment should give design and test personnel a commonality of purpose. We've needed that for a long time.

It appears that the military customer is finally saying to his suppliers, "You know more than I do about what I need." Earlier documents created an adversary relationship between military customer and contractor. Hopefully, 810D will encourage cognizant individuals on both sides to talk directly about what is needed, without going through several layers of management.

Environmental tests are to be based upon environmental measurements, rather than (as before) upon classical test curves. The latter are still available, but as "fall back" or "default" values. They are the last resort, to be employed only if, for some reason, real data is not available. They are conservative, so everyone benefits from tailoring.

The vibration and shock tests of 810D will be difficult to apply in laboratories that lack computer control. Even with computers, some Methods will not immediately be possible due to lack of firmware from vendors. Hopefully, that situation will ease within a year.

Table of Contents (The Table of Contents is expanded at the beginning of each Section)

vi

Wayne Tustin is widely known as "Mr. Random Vibration." He received his BSEE from the University of Washington, Seattle, in 1944. His introduction (unprepared, as is common among his students today) to vibration was at Boeing, 1948-53. He directed field service and technical training, 1954-61, at MB Electronics, New Haven, Connecticut, a pioneer manufacturer of shakers for random vibration testing. 1962 saw the founding of Tustin Institute of Technology (TIT) at Santa Barbara, California. TIT is a private engineering school, emphasizing vibration (particularly random) and shock (also intense noise) testing, measurement, analysis and calibration. He trains groups at their facilities. He also teaches a few "open" courses at Santa Barbara and elsewhere around the world. Finally, some individuals study with him via the mails. He is a Fellow of the IES and a member of SEE, ASME and ASQC and a licensed Professional Engineer—Quality in the State of California. He has published literally hundreds of articles (mainly in the USA) and several texts.

Robert Mercado has been involved in environmental testing since 1968, when he joined Dayton T. Brown, Inc., a prominent New York commercial testing laboratory. There he specialized in all phases of dynamic testing, including vibration, shock, acceleration and seismic; also calibration and maintenance of test equipment. In 1979 he relocated to California, joining Santa Barbara Research Center (a subsidiary of Hughes Aircraft) as a Quality Assurance Engineer. Here he specialized in environmental testing and stress screening, using electromagnetic and pneumatic (quasi-random) techniques. In 1983, he joined Screening Systems, Inc., of Irvine, California. Mr. Mercado has published several technical articles on stress screening, in the USA and abroad.

Introduction

Few environments in which military and commercial equipment operate include purely sinusoidal vibration — random or near-random vibration is the norm in this world we live in. But for many years, we designed equipment to operate in sinusoidal vibration environments, and it should have come as no surprise when this equipment failed in vibration much more often than we had expected.

Now we know better, and industry is pocketing the profits of using random vibration in their factory test programs to more nearly *simulate* the real environment of the equipment they are developing. The test equipment costs more but the payoff is even greater. Designs are emerging which can tolerate without failure the vibration stresses they will experience in service.

During the space programs of the 1960s, we learned another valuable lesson about random vibration. Especially in conjunction with thermal cycling, the right spectrum of random vibration could *stimulate* latent manufacturing defects in the factory. In recent years, 90% of the failures of Navy equipment in the Fleet have been due to manufacturing defects, not design problems. Furthermore, I have visited contractor plants where over 50% of their production was in rework — in what I call their "hidden factory."

So I transferred some space technology to the Navy when I published NAVMAT Publication P-9492 on environmental stress screening. Since then, more and more Navy contractors have invested in the necessary equipment and have discovered for themselves the early payback and additional profits of eliminating the hidden factory, while the Navy is finally beginning to get some really defect-free equipment in the Fleet.

I urge you to take Wayne's lessons to heart. You can only profit from their application.

Willis J. Willoughby, Jr.
Deputy Chief of Naval Material (RM&QA)
Naval Material Command
Washington, D.C.
February 13, 1984

Section 0
Sources of Vibratory Energy in Machines and Vehicles

0.1 Your automobile provides vibration (and shock) examples. Your automobile can introduce you to vibration measurement and analysis. How is vibration described? What are the sources? Consider the engine of Fig. 0.1. How shall we describe its vibration?

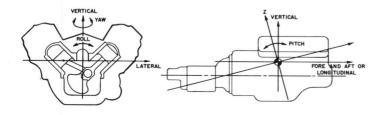

Fig. 0.1 An automobile engine generates a variety of vibrations.

0.1.1 Rotational unbalance. Flexible supports (engine mounts) isolate the engine from the frame. But these mounts are not very effective when the engine idles. Motions and forces are coupled mechanically through a structure. (You also get acoustically-coupled energy called sound or noise.)

What causes that vibration? Sec. 8 discusses the many sources. For now, visualize the rotating crankshaft and flywheel not perfectly balanced about the centerline. A whirling, eccentric force results, passes through the crankshaft bearings and causes the whole engine to vibrate. At idling speed, we can see large displacements.

Unbalance vibration occurs at a certain *frequency*, the number of events per minute or per second. To state frequency in cps or the SI term, hertz (abbreviated Hz), divide RPM by 60. Thus an unbalanced rotor running at 1,800 RPM generates a frequency of 1,800 cpm or 30 Hz.

Sec. 1 further discusses the ideas of natural frequency f_n (frequency at which a structure readily vibrates) and of resonance (exciting force f_f is at, or very close to, f_n). At certain RPMs there is much vibration. Avoid operating very long at those *critical speeds*.

This simplified discussion introduces technical terms you will need. The vibration forces of your engine occur at several frequencies; the most common is at "one times RPM" or first-order. Others may be more damaging. When *any* exciting or forcing frequency f_f coincides with engine or supporting structure f_ns we have the condition of resonance. Your automobile frame is fairly stiff; that is, its natural frequencies are fairly high. By placing a relatively soft *vibration isolator* or *engine mount* between engine and frame, you *isolate* some engine vibrations and avoid many frame resonances; see Sec. 0.1.4, 0.2.12 and 1.1.5. Expect large engine motions, per Fig. 0.1, especially when idling.

0.1.2 Bearings per Fig. 0.2 introduce some vibration. Rotation moves the bearing supports. With a stethoscope you could hear the sound. To measure and analyze the motions, you would use a vibration pickup, or sensor, or *transducer*, to produce an electrical signal for a meter and analyzer. We will discuss vibration measurement in Sec. 3-6 and vibration analysis in Sec. 7 and 8.

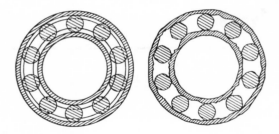

Fig. 0.2 A representation of smooth and rough ball or roller bearings. Courtesy Bendix.

Automobile mechanics are satisfied with bearings which spin easily (no audible noise, minimal friction). But submarine bearings must be extremely smooth. Transmitted through the hull, vibration interferes with sonar operations and radiates as sound (revealing the submarine's position, course and speed). One can select quiet bearings on the Anderometer of Fig. 0.3, using vibration sensors, electronic amplification, a loudspeaker for listening and meters for readout.

Fig. 0.3 Anderometer for classifying bearings as to smoothness.

0.1.3 Cooling fan. Fan rotational unbalance forces are minimized by balancing. Blades flutter, generating noise and vibration. As they sweep past obstructions (briefly impede air flow), pressure pulsations generate forces and noise. Duct or shroud resonances may be excited, either by mechanically-transmitted vibration or by acoustical (air) coupling.

0.1.4 Water pump. Vibration and noise come from gear mesh, bearings and intermittent or pulsating water flow. Cavitation might occur. If any excitation coincides with f_{ns} of pump or support, resonance will magnify the effects and contribute to overall vibration and noise.

0.1.5 Electrical generator or alternator. Possible sources of vibration and noise: electrically-generated forces, rotational unbalance, belt noise, faulty bearings and internal cooling fan.

0.1.6 Gears within the engine and in the transmission and differential can produce vibration and noise. One possible f_f: multiply gear RPM by the number of teeth. Faulty teeth, per Fig. 0.4, will generate large forces.

Fig. 0.4 Faulty gears create intense noise and vibration and may fail catastrophically.

0.1.7 Road surface inputs affect the wheels and suspension. We will discuss *random* vibration in Sec. 21 and 22. Sometimes your car receives a more intense, brief *shock* input per Sec. 32.

0.2 Wide variety of vibration problems. Readers may be intrigued at the tremendous variety of dynamics problems that trouble various industries.

In vital installations, machine vibration should be regularly checked. If a check per Sec. 8 shows a large vibration increase, repairs can prevent catastrophe. Sec. 7 explains electronic analysis to detail any machinery, transportation, service or laboratory vibration spectrum. An abnormally-large vibration at a particular frequency diagnoses which part is faulty.

Noise and vibration often warn of impending failure. We want to avoid breakdowns. We react to long-continued excessive noise and vibration with tiredness or fatigue. Similarly, machines wear out faster under repeated high stresses, and may fail. Parts loosen. Bearing temperatures rise and may "burn out." These ideas apply to all vehicles: trucks, ships, airplanes; to all machinery.

0.2.1 Office vibration. Modern office buildings feature little mass and little structural damping; see Sec. 1.1. Powerful blowers, pumps, compressors, etc. are often installed on roofs. The upper floors (which should bring high rents) are sometimes vacant due to vibration and sound. Fans and other mechanisms inside office machinery can cause subtle malfunctions and fatigue failures. See "The Business Machine Vibration Environment," by Skinner and Zable of IBM, in THE JOURNAL OF ENVIRONMENTAL SCIENCES, Sept/Oct 1978.

0.2.2 Wind forces on electrical transmission lines create vibration, eventually fatiguing lines and supports. Tall building sway causes personnel discomfort.

0.2.3 Compressor stations on gas pipe lines, unattended, may be 100 or 200 miles from maintenance personnel. Remotely monitored signals warn of imminent bearing or gear failure or of "thrown" turbine blades.

0.2.4 Vibration measurements as a quality control tool. As a "quick check" on bearing quality, shaft alignment and dynamic balance, the inspector of Fig. 8.15 will reject any units that exceed some upper limit on vibration.

0.2.5 Aircraft and missile tests. Vibration tests on aircraft parts prove units will function in spite of vibration. Environmental vibration tests are described in Sec. 10 though 27.

Fig. 0.5 Commode for passenger aircraft being vibration tested. Courtesy Inland Testing Labs.

0.2.6 Machinery. Here vibration causes chatter, poor finishes and excessive rejects. Down time and maintenance costs can skyrocket.

Machine tools users often find that isolation per Sec. 0.2.12 and 1.1.5 against vibration and shock is helpful. Cold-header bolt-making machines, lagged down to concrete floors, tear loose due to severe shocks. Rather, float machines on rubber mounts which deflect slightly each stroke, dissipating energy.

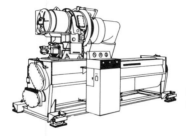

Fig. 0.6 Machine tool isolated from factory floor.

Weaving looms were once bolted into soft wooden buildings. Shuttle acceleration and deceleration forces were dissipated by building motion. However, when looms were moved to concrete-floored buildings, maintenance costs soared. Flexibility was restored by floating looms on soft pads.

0.2.7 Electronics production stress screening, formerly called "burn in," employs temperature shock and random vibration. The goal of screening such subassemblies as printed circuit boards is to winnow out units having manufacturing defects (such as poorly soldered connections or contaminants). See Sec. 30.

0.2.8 Transportation damage. The vibration and shock of transport costs billions of dollars annually. Rough factory handling is followed by truck or rail travel, then rough handling upon receipt. Amazingly little is *known* about these environments. See "An Assessment of the Common Carrier Shipping Environment," by Ostrem and Godshall, 1979, Forest Products Laboratory, U. S. Dept. of Agriculture, Madison, WI; see also pg. 83, Part 2, SHOCK AND VIBRATION BULLETIN 50. See Sec. 1.1.5 re isolation against such damage and Sec. 33.5 re testing of packages.

0.2.9 Loosening of fasteners. Bolts, etc. loosen under vibration and shock. Friction elements in certain fasteners are helpful. Recent bolt studies have shown the importance of bolt stiffness, part stiffness and preload.

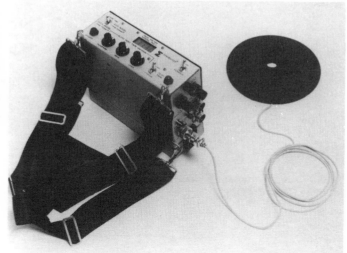

Fig. 0.7 Ride Meter. Courtesy Endevco.

0.2.10 Vibration and shock effects upon humans. Sustained vibration at certain frequencies annoys, and sometimes physically damages people. An example: a truck driver's painful back can eventually incapacitate him. Fig. 0.7 shows a special seat pad containing single or multiple axis (see also Sec. 5.5) accelerometers. These are connected to an electronics package worn by the test subject. Electronic filtering (see also Sec. 7.4) emphasizes motions at most damaging frequencies. The numerical readout helps to rate vehicles and seats.

Vibration White Finger (VWF) industrial disease, (Raynaud's Syndrome) involves long exposure to severe hand vibration. See NIOSH (Cincinnati, Ohio) Technical Report by D. Wasserman, 1982, videotape #177.

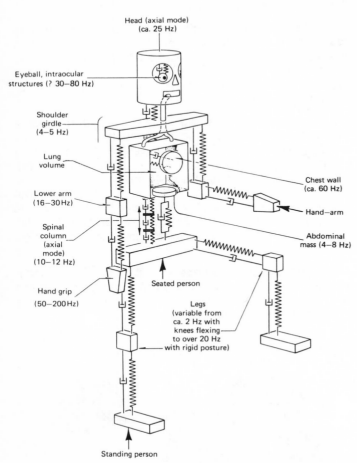

Head (axial mode)
(ca. 25 Hz)

Eyeball, intraocular
structures (? 30–80 Hz)

Shoulder
girdle
(4–5 Hz)

Lung
volume

Lower arm
(16–30 Hz)

Spinal
column
(axial
mode)
(10–12 Hz)

Hand grip
(50–200 Hz)

Chest wall
(ca. 60 Hz)

Hand–arm

Abdominal
mass (4–8 Hz)

Seated person

Legs
(variable from
ca. 2 Hz with
knees flexing
to over 20 Hz
with rigid posture)

Standing person

**Fig. 0.8 Complex dynamic model of human (standing).
Courtesy Bruel & Kjaer.**

Sustained very low frequency sounds and/or vibrations can cause illness. Little data has been scientifically gathered, and no standards exist. (See AGARD monograph No. 151 "Aeromedical Aspects of Vibration and Noise" by Guignard and King, NATO 1972. Available from the NTIS as AD-754631.)

0.2.11 Vibrating conveyors for moving bulk materials are either (1) brute force, with a powerful motor, or (2) resonant, with the drive system operating at a mass/spring resonance, requiring a relatively small motor. See "Vibrating Conveyor and Feeder Systems," by W. L. Hickerson, in MECHANICAL ENGINEERING of December, 1967.

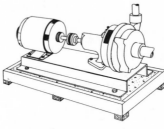

Fig. 0.9 Building services pump isolated from building structure.

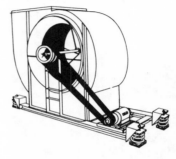

Fig. 0.10 Air-circulating fan isolated from building structure.

0.2.12 Vibration isolation. Fig. 0.6, 0.9 and 0.10 show isolated machines. Precision machine tools should be protected from building vibration (caused by other machines or by traffic). Machines should be permitted some motion relative to their supports, to reduce internal vibration or to keep their vibration from exciting building resonances. See also Sec. 1.1.5.

Section 1
Introduction to
Dynamic Motion

1.0 Solid-body motion vs. deformation. What vibrations (motions) might the unit of Fig. 1.1 have? One is called whole-body: the entire assembly and foundation moves as a unit, bobbing and rocking on its spring supports. Another is deformation, relative motion between various parts, as in the bending of a ship's hull at sea.

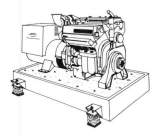

Fig. 1.1 Engine-driven generator on isolators.

Certain whole-body motions and certain deformations are most common: those easiest to excite. We speak of these occurring at certain *natural frequencies*. When exciting (forcing) frequencies coincide with any of the natural frequencies we get relatively severe vibratory responses. This condition is called *resonance*. In Sec. 2 we will quantize motions, deformations and forces.

In Sec. 1 we examine single-axis vibration, while in Sec. 29 we consider torsional or angular vibration equivalents.

1.1 Dynamic motion of simple spring-mass system. Consider the simple mechanical system shown in Fig. 1.2. It includes a weight W (mass M is technically preferred), a spring of stiffness K and a viscous damper C. Spring rate K is expressed in newtons/mm or lb/in. A static force will statically deflect the spring, increasing its length from l to $l + \delta$. The mass is constrained to just one motion — vertical translation. True single-degree-of-freedom (SDoF) systems don't exist, but many systems' dynamic behaviors approximate SDoF behavior over small ranges of frequency.

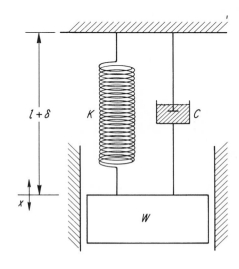

Fig. 1.2 Single Degree of Freedom (SDoF) model.

1.1.1 Free vibration and natural frequency. Mentally pull weight W down a short distance further and let it go. W will oscillate (free vibration) at natural frequency f_n, expressed in cycles per second (cps) or in hertz (Hz). We here ignore the effect of damper C. Like the "shock absorbers" or dampers on your automobile's suspension, C dissipates vibratory energy as heat; oscillations die out. f_n may be calculated by this formula:

$$f_n = \frac{1}{2\pi} \sqrt{\frac{Kg}{W}} \qquad (1.1)$$

It is often convenient to relate f_n to the static deflection δ caused by Earth's gravity. $F = W = Mg$, where $g = 386$ in/sec^2 $= 9807$ mm/sec^2, opposed by spring stiffness K. On the moon g, W and δ are about 1/6 as large as on Earth. Yet f_n is the same.

$$f_n = \frac{1}{2\pi} \sqrt{\frac{K}{M}} \qquad (1.2)$$

Assume an increasing weight W (increasing mass M) on a spring of stiffness K. Table 1.1 shows how δ increases and natural frequency f_n drops.

Note that f_n depends upon δ. If load and stiffness change proportionately, f_n does not change.

Peak-to-peak or Double amplitude (D) of vibratory displacement remains constant if no damping uses up energy. The spring's potential energy becomes zero as the mass passes through its original position and becomes maximum at each displacement extreme. Kinetic energy is maximum as the mass passes through zero (greatest velocity) and is zero at each displacement extreme (zero velocity). Without damping,

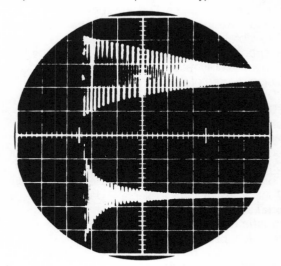

Fig. 1.3 Decaying sinusoidal vibration after SDoF system is plucked.

"USA" SYSTEM

$$\delta = \frac{F}{K} = \frac{W}{K}$$

$$f_n = \frac{1}{2\pi} \sqrt{\frac{g}{\delta}} = \frac{1}{2\pi} \sqrt{\frac{386}{\delta}}$$

$$= \frac{19.7}{2\pi\sqrt{\delta}} = \frac{3.13}{\sqrt{\delta}} \qquad (1.3)$$

INTERNATIONAL SYSTEM

$$\delta = \frac{F}{K} = \frac{Mg}{K}$$

$$f_n = \frac{1}{2\pi} \sqrt{\frac{g}{\delta}} = \frac{1}{2\pi} \sqrt{\frac{9807}{\delta}}$$

$$= \frac{99.1}{2\pi\sqrt{\delta}} = \frac{15.76}{\sqrt{\delta}} \qquad (1.4)$$

TABLE 1.1

Let K = 1,000 lb/in and vary W			Let K = 1,000 N/mm and vary M		
W, lb.	$\delta = \frac{W}{K}$, in.	f_n , Hz	M, kg.	$\delta = \frac{9.81M}{K}$	f_n , Hz
.001	.000001	3130	.00102	10 nm	4980
.01	.00001	990	.0102	100 nm	1576
.1	.0001	313	.102	1 μm	498
1.	.001	99	1.02	10 μm	157.6
10.	.01	31.3	10.2	100 μm	49.8
100.	.1	9.9	102.	1 mm	15.76
1000.	1.	3.13	1020.	10 mm	4.98
10000.	10.	.99	102000.	1 m	.498

energy is continually shifted from potential to kinetic energy. However, in practical systems some damping C gradually decreases motion; energy is converted to heat. A vibration sensor on weight W would give oscilloscope time history patterns like Fig. 1.3. The lower pattern represents more damping; motion decreased more rapidly.

1.1.2 Response to forced vibration - resonance. Let us now imagine that the overhead "support" of Fig. 1.2 is being vibrated by a shaker. D is constant at 1 inch or 25 mm. Shaker frequency varies. How much "response" vibration occurs at weight W? The answer depends upon

(1) the frequency of "input" vibration, and

(2) the natural frequency and damping of the system.

Visualize an f_n of 1 Hz and a forcing frequency of 0.1 Hz, 1/10 the natural frequency. Weight W has about the same motion as does the input, 1 inch or 25 mm D.

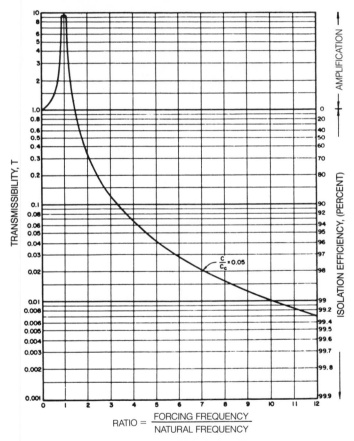

Fig. 1.4 Transmissibility graph for SDoF system having 5% critical damping. Courtesy Consolidated Kinetics.

1.1.3 Transmissibility (Tr) ratio. Fig. 1.4 is a graph of transmissibility (Tr) vs. frequency, which, for an undamped system, is expressed as:

$$Tr = \frac{1}{1 - \left(\frac{f_f}{f_n}\right)^2} \quad (1.5)$$

Tr represents response magnitude relative to input magnitude

(two displacements, two velocities, two accelerations, or two forces), per

$$\frac{\text{response magnitude}}{\text{input magnitude}} \quad (1.6)$$

Visualize a shaker input, $f_f = 0.1$ Hz, into an SDoF system having $f_n = 1.0$ Hz. Shaker D and mass W's D are equal. Tr = 1. Find this situation at the left edge of Fig. 1.4.

1.1.4 Single degree of freedom system - resonance. As we increase the forcing frequency, we find that the response increases. How much? It depends upon damper C. Let us assume that C is 0.05 critical. When f_f reaches 1 Hz (exactly f_n) weight W has a response D of about 10 inches, 10 times as great as D_{in}. At this "maximum response" frequency, we have the condition of "resonance": $f_f = f_n$. The transmissibility ratio Tr = 10. We can use the phrase "Q = 10".

As we increase f_f from this point (see Fig. 1.4), response decreases. At $f_f = 1.414$ Hz, the response D_{out} is again 1 inch, and Tr = 1 again. As we further increase f_f, the response decreases further. At $f_f = 2$ Hz, $D_{out} = 0.3$ inch. At $f_f = 3$ Hz, $D_{out} = 0.1$ inch. Find these points on Fig. 1.4.

Note that the abscissa of Fig. 1.4 is "normalized;" that is, the Tr values would be the same for another system whose f_n is 10 Hz, when f_f is respectively 1, 10, 14.14, 20 and 30 Hz.

1.1.5 SDoF system - isolation. The region above about 1.414 f_n (where Tr is less than 1) is called the region of "isolation." Weight W has less vibration than the input; W is isolated. Vibration isolators (rubber elements or springs) reduce vibration inputs to delicate units on aircraft, missiles, ships, etc. We normally set f_n (by selecting isolator K) well below the expected f_f. Thus if our support vibrates at 50 Hz, we might select isolators whose K makes f_n 25 Hz or less. According to Fig. 1.4, we will have best isolation at 50 Hz if f_n is as low as possible. (However, we don't use too-soft isolators because of difficulties from too-large static deflections, and need for large clearance to nearby structures.) See Sec. 32.8 on shock isolation.

Your automobile suspension provides an example. Road inputs well above suspension f_n (due to driving at high speed) have little effect. But at low speed you feel every bump. A luxury automobile has a larger δ (lower f_n) and better isolates you from road inputs, at whatever speed. Note that you must have relative motion (between wheel and chassis) to achieve isolation.

Imagine a weight supported by a spring whose stiffness K is chosen for $f_n = 10$ Hz. At $f_f = 50$ Hz, the frequency ratio will be 50/10 or 5, and we can read Tr = 0.042 from Fig. 1.4. The weight would "feel" only 4.2% as much vibration as if it were rigidly mounted to the support. We might also read the "isolation efficiency" as being 96%.

1.1.6 SDoF system - effect of damping. If our 50 Hz vibration source were a motor, how did it get "up to speed?" If it passed slowly through 10 Hz the isolated item would at resonance experience about 10 times as much vibration as if rigidly attached, without isolators. Damping helps limit the "Q" or "mechanical buildup" at resonance. This extends structural life before fatigue failure occurs.

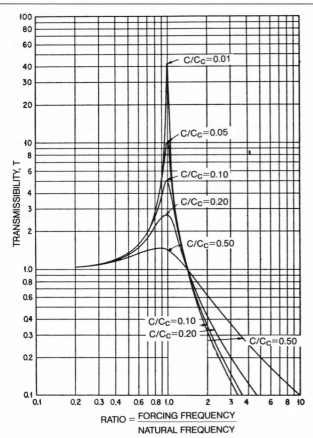

Fig. 1.6 shows the shift in frequency ratio that results from various amounts of damping.

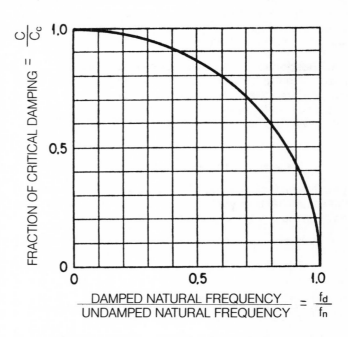

Fig. 1.6 Effect of damping on SDoF system. After Morrow.

RATIO = $\dfrac{\text{FORCING FREQUENCY}}{\text{NATURAL FREQUENCY}}$

Fig. 1.5 Transmissibility graph for SDoF system having various fractions of critical damping. After Morrow.

Observe Fig. 1.5, plotted for several different C values. With little damping, resonant magnification is high. With more C, Q or maximum transmissibility is reduced. For instance, where C/C_c is 0.01, "Q" is about 40. C_c is a very large, somewhat theoretical amount of friction or damping (can be computed by $2\sqrt{KM}$) which prevents oscillation or "ringing" after a disturbance. (There is an exact electrical analogy, resistance,which keeps electrical circuits from oscillating after a disturbance.)

Even higher Q values (to over 2,000) are found with light damping. Most structures (ships, aircraft, missiles, etc.) have Q's ranging from 10 to 40. Bonded rubber vibration isolating systems often have Q's around 10. If additional damping is needed (to keep Q lower), dashpots such as your car's "shock absorbers" or other damping devices can be used with SDoF systems. There will be less buildup at resonance but isolation is not as effective.

Large amounts of damping can significantly lower the natural frequency from that of Equation 1.5 to

$$Tr = \left[\frac{1 + \left(\dfrac{f_f}{f_n}\right)^2 \left(\dfrac{2c}{c_c}\right)^2}{\left(1 - \dfrac{f_f^2}{f_n^2}\right)^2 + \left(\dfrac{f_f}{f_n}\right)^2 \left(\dfrac{2c}{c_c}\right)^2} \right]^{\frac{1}{2}} \quad (1.7)$$

Fig. 1.5's damping was viscous. Somewhat different curves result from Coulomb or hysteretic damping; these more accurately model the behavior of commercial isolators, depending upon use of natural vs. synthetic rubber and whether the elastomer is loaded in shear, in compression or tension, or in torsion. Viscous damping is usually assumed, however, for simplicity. And because curves similar to Fig. 1.5 are readily available.

We're only introducing the subject of vibration isolation. We have only discussed, for example, isolating SDoF systems. Considerable specialized experience and training are needed for selecting "real world" isolators and for properly locating them.

Vibration isolators (Sec. 1.1.5) take many forms. Natural or synthetic rubber is temperature limited, but is used under machine tools and even for isolating buildings from earthquakes.

Fig. 1.7 Wire rope vibration isolators. Courtesy Aeroflex.

Ordinary springs have little damping, with dangerously high peak transmissibility at resonance. Friction must be added. Fig. 1.7, for instance, shows a stainless steel cable wound into a helix so individual wire strands rub. Alternately (Fig. 1.8) stainless steel wire mesh can be placed inside a helical spring.

Fig. 1.8 Typical helical spring isolator with wire mesh for damping. Courtesy Barry Controls.

For extremely low f_n (very large δ) air columns, as in air-ride busses and trucks, are useful.

1.2 Multiple (mechanical) degrees of freedom. Fig. 1.2 shows guides that constrain W's motion to up-and-down translation. One displacement measurement describes instantaneous arrangement of its parts. Another SDoF system: a wheel on a shaft. Twist the wheel; it will oscillate at a certain f_n, determined by shaft stiffness and wheel inertia. This rotational system is an exact counterpart of the SDoF system shown in Fig. 1.2. See Sec. 29.2. Without the guides of Fig. 1.2, W could have five additional degrees of freedom. Visualize the six:

Vertical translation,
North-South translation,
East-West translation,
Rotation about the vertical axis
Rotation about the North-South axis, and
Rotation about the East-West axis.

Describing the various whole-body motions that could occur would require six measurements.

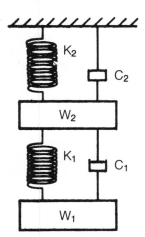

Fig. 1.9 Multiple Degree of Freedom (MDoF) system.

Attach the system of Fig. 1.2 to another mass, spring and damper, per Fig. 1.9. This is more typical of many actual systems. Machine tools, for example, don't attach directly to bedrock, but rather to other structures with their own vibration characteristics. Describing all possible solid-body motions of W_1 and W_2 would require 12 measurements.

1.3 Continuous systems. Extend that reasoning. Include additional masses, springs and dampers, additional mechanical degrees of freedom — possible motions. Finally, consider a continuous beam or plate, where mass, spring and damping are distributed (rather than being concentrated as in Figs. 1.2 and 1.9). Visualize an infinite number of possible motions, an infinite number of f_ns.

1.3.1 Continuous systems - single frequency excitation. Three photos of Fig. 1.10 were taken at three different forcing frequencies. Vertical shaker motion drives two beams. (For now, only consider the left-hand solid beam.) Frequency is adjusted first to excite the fundamental shape (also called the "first mode" or "fundamental mode") of beam response. Then frequency is increased to excite the second and third modes; the points of minimum D are called "nodes" and the points of maximum D are called "antinodes." Salt sprinkled on the beam will collect at the nodes. With a strobe light and/or vibration sensors we could show that (in a pure mode) all points move either in-phase or out-of-phase; phase reverses from one side of a node to the other. Also that bending occurs at the attachment point and at the antinodes; here is where fatigue failures usually occur. We'll have an infinite number of such resonances between zero and infinity f_f. Mentally extend this reasoning to vehicles, machine tools, appliances and various structures; an infinite number of resonances can exist. Fortunately, the exciting frequencies are limited. But whenever

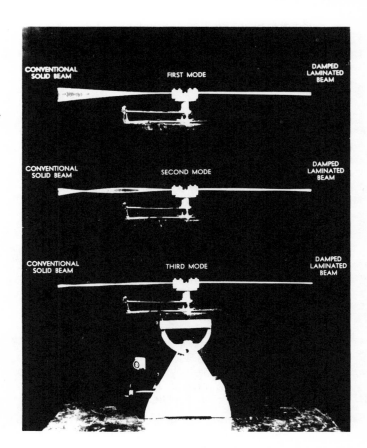

Fig. 1.10 Response of undamped (left) and damped cantilever beams. Courtesy Barry Controls.

an f_f coincides with one of the f_ns, we will have resonance: high stresses, forces and motions.

All remarks about free vibration, resonance and damping apply to continuous systems. "Pluck" the left hand beam of Fig. 1.10. It will respond in one or more of the patterns shown. We know that its vibration will gradually die out per the upper trace of Fig. 1.3, due to internal (or hysteresis) damping. Stress reversals create heat, using up vibratory energy.

Vibration dies out faster, per the lower trace of Fig. 1.3, on the right-hand "sandwich" cantilever beam. Here a viscoelastic layer joins two metal layers, per Fig. 1.11. Stress reversals now use up vibratory energy faster. Free vibration dies out more quickly.

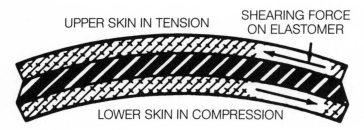

UPPER SKIN IN TENSION   **SHEARING FORCE ON ELASTOMER**

LOWER SKIN IN COMPRESSION

Fig. 1.11 Viscoelastic damper constrained between two metal layers. Courtesy Lord.

The statements we made earlier about forced vibration and resonance on SDoF systems also apply to continuous systems. At very low forcing frequencies, the response motion is the same (Tr = 1) all along the beams. At certain f_n's, large motion results, especially on the solid beam.

1.3.2 Critical frequencies are those at which continued operation *can* be dangerous. Q (and the degree of danger) depend upon the damping present, per Sec. 1.1.6. Often the first resonance is the most critical. Consider a printed circuit board (PCB). Maximum dynamic displacement, maximum bending stresses in the board, also maximum stresses on attached components, usually occur at the lowest (first) resonance. Stiffening the PCB (making δ less) will raise the first f_n and reduce stresses. Thicken the PCB, reduce its dimensions or attach a beam or other structural element.

In planning a resonance search, someone may feel that resonant magnification Q less than, say, 2 should be ignored. Resonances still exist, but he decides that they are not critical, should not be reported.

1.3.3 Continuous systems—multi-frequency excitation. Several response modes can occur at once, if several forcing frequencies are present, as with complex vibration or broadband random vibration (see Secs. 7 and 21). Several modes can simultaneously be excited by a shock pulse (and then die out). Certain modes are most likely to cause failure; the f_f s causing these are called "critical frequencies."

1.3.4 Continuous systems—effects of damping. The lengths of the two beams of Fig. 1.10 were adjusted until their f_ns were identical. The conventional solid beam responds more strongly —its maximum transmissibility "Q" is greater. Which beam will first fail in fatigue? We add damping to extend fatigue life.

1.4 Modal analysis is generally beyond the scope of this text, but readers will encounter the term elsewhere. The subject may be broken down into two main areas: (1) Theoretical analysis, in which natural frequencies, magnifications and detailed mode shapes (as in Fig. 1.9 for a very simple structure) and other data (such as stresses) are predicted. These are based upon dimensions and material properties of a structure that will be built. (2) Experimental analysis on a structure, to (hopefully) verify the theoretical analysis and to demonstrate natural frequencies, magnifications and mode shapes.

Experiments are performed on prototype automobiles, aircraft, machine tools, etc. Slow single-frequency analog controlled sine forcing was once popular, but is now seldom used. Under digital computer control, and in order to excite all modes simultaneously, broad-spectrum random forcing is much faster. Simultaneous inputs are obtained from several shakers. Shock forcing (fastest) can be used (see Sec. 32.7.) In any event, the applied force is measured (see Sec. 6.1), along with the resulting motions. Measurements are today computer processed to give the needed information about the structure's modal responses.

1.5 Acoustic responses. Intense (above about 130 dB) sounds can cause vibration modes to be excited, especially in thin panels; stresses in aircraft and missile skins (also on printed circuit cards in electronics units) may cause fatigue and other failures. Damping treatments are often very effective in reducing such vibration. See Sec. 31.

1.6 Dynamic vibration absorber. A resonant vibration absorber can reduce motion, if vibration is at a fixed frequency. Imagine, for example, weaving machines in a relatively soft, multi-story factory building. They happen to excite a floor resonant motion. A remedy: attach pails via spring supports under those floors. Fill the pails with sand until f_ns = f_f. A dramatic improvement results. Any change in exciting frequency necessitates a bothersome readjustment. Torsional absorbers are discussed in Sec. 29.1.

1.7 Resonance can be helpful. Resonance is sometimes helpful and desirable; at other times it is harmful. Some readers will be familiar with deliberate applications of vibration to move bulk materials (as mentioned in Sec. 0.2.12), to compact materials, to remove entrapped gases or to perform fatigue tests. Maximum vibration is achieved (assuming vibratory force is limited) by operating a system at resonance.

1.8 Suggestions for designers. Generally, vibration inputs (single or multiple frequencies — a line spectrum) to an equipment will vary somewhat in frequency (as RPM of an engine varies, for example), or they will occur simultaneously at all frequencies, as with random vibration—see Sec. 21. Designers try to avoid resonances. They will try to place all f_ns well above the highest f_f they expect in service or in test. With reference to the transmissibility curves of Figs. 1.4 and 1.5, designers generally prefer the left-hand region, where Tr ‹ 1. This may not be possible; getting f_ns above the highest f_f may add too much weight. What then?

Sec. 1.1.5 suggests supporting items on soft elements such as those in Figs. 1.7 and 1.8. Since those elements, when supporting the equipment, have low natural frequencies, operation is now in the right-hand "isolation" region of the transmissibility graph. Suppliers of isolators (Aeroflex, Barry, Lord, for example) offer assistance in selecting and in proper placement of isolators.

We can add structural damping after fabrication. We can bond sheets of viscoelastic material onto PCB's. Damping not only reduces resonant response motion, extending PCB fatigue life, it also reduces the environmental vibration "input" to delicate parts on the board. If you cannot isolate a delicate unit, reduce the unit's input vibration by adding damping to the supporting member. Viscoelastic materials can be brushed or sprayed (as in undercoating automobiles), or bonded onto ship hulls, outboard motor housings, typewriter and computer housings, etc.

Alternately, fabricate circuitry onto a damped card, rather than on ordinary fiberglass board material; per the broken graph of Fig. 1.12, at 150 Hz and at 365 Hz input motion is greatly magnified. The solid line shows the lowered response of a damped board, reducing the vibration input to components. (Note that adding damping lowers f_ns somewhat.)

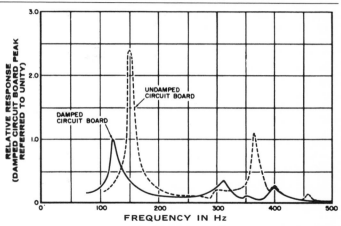

Fig. 1.12 Responses of damped vs. undamped circuit boards. Courtesy Barry Controls.

1.9 Closing. This material on sinusoidal vibration, resonance and damping is greatly simplified. It will prepare those students who wish to study further, using the many excellent textbooks that are available. An example: "Vibration Control," by MacDuff and Curreri, McGraw-Hill, 1958.

Fig. 1.13 Tacoma Narrows Bridge, Washington State, November 7, 1940, shortly before wind-induced vibrations destroyed the structure. Photo courtesy Farquharson Collection, University of Washington.

Section 2
The Specialized
Language of Vibration

2.0 Terms. "I often say that when you can measure what you are speaking about, and express it in numbers, you know something about it; but when you cannot express it in numbers, your knowledge is of a meager and unsatisfactory kind; it may be the beginning of knowledge, but you have scarcely, in your thoughts, advanced to the stage of *science*, whatever the matter may be." William Thomson, Lord Kelvin, "Popular Lectures and Addresses," 1891-1894.

To communicate, you must speak the right language. Even then, translating idioms is very difficult. You've heard a foreigner having trouble communicating. No matter that he has "studied English." The words may not mean what he *thinks* they mean. "She is a sight" is not the same as "She is a vision."

Before a designer, a quality engineer or a reliability specialist tries to communicate with environmental test personnel, he must speak laboratory language as well as his own.

Vibration test, instrumentation and metrology people use the words of Table 2.1:

Table 2.1

Parameters	Symbol for Instantaneous Value	Symbol for Most Common Statistical Value	Most Common Statistical Value
Displacement	x	D	Peak-to-peak
Velocity	v	V	Zero-to-peak
Acceleration	a	A	Zero-to-peak

In the classroom, with sinusoidal vibration (all vibratory motion occurs at a single frequency f), the three quantities relate simply. (Note that "real world" vibration is seldom sinusoidal. See Sec. 7 and 8 on complex vibration and Sec. 21 on random vibration.) Here in Sec. 2 we pretend that vibration is sinusoidal; but remember that sinusoidal vibration is seldom found outside of textbooks, older testing specifications and

calibration laboratories. Sec. 2 deals with single axis vibration (also quite rare — see Sec. 18 on multi-axis vibration and Sec. 29 on torsional vibration).

2.1 Frequency f is the number of occurrences in one minute or one second. If one cycle takes 0.1 sec, f = 10 hertz (abbreviated Hz), meaning 10 cycles per sec.

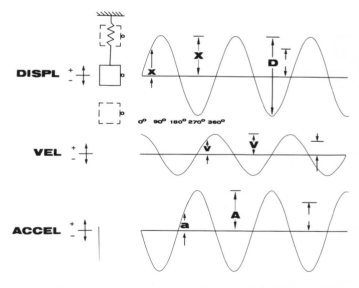

Fig. 2.1 Relationships between sinusoidal instantaneous x, v or ẋ and a or ẍ. Peak statistical values are dimensioned and identified. RMS statistical values are dimensioned only.

2.2 Displacement, to a schoolboy, might involve automobile motion along a road, with distance in kilometers or miles. That is overly simplified; "vibration" suggests an oscillating, varying, reversing displacement. Consider some element moving sinusoidally per Fig. 2.1 (upper graph). The instantaneous position or displacement x varies sinusoidally with time. With calculus, we can instead write

$$x = Xsin2\pi ft \qquad (2.1)$$
where f is frequency in Hz, and
t is time in seconds.

We seldom use x but we often quantize the maximum or peak displacement X. Our concern may be vibrating members exceeding some clearance or bending stress limit.

Per Fig. 2.1, at what phase angles does x equal X? −X?

2.3 Velocity. The schoolboy's idea of velocity: distance traveled per unit time, km or miles per hour. Our concept of vibratory velocity per Fig. 2.1 (center) involves varying magnitude and reversals. Instantaneous velocity v at any time t represents the x curve's slope, which changes from + to − to + each cycle. When that slope is zero, v is zero. When x is zero, its slope is greatest (either + or −) and v is maximum (either + or −). Measure that slope at additional points to fill in the v curve. Note that v leads x by 90°.

Alternately, let us develop an equation for v. We "differentiate" (find the slope of) Eq. 2.1 above:

$$v = 2\pi fXcos2\pi ft \qquad (2.2)$$

In practice, nearly all metrologists prefer D, rather than X. Since D = 2X, Eq. 2.2 becomes

$$v = \dot{x} = \pi fDcos2\pi ft \qquad (2.3)$$

We seldom use v. But we may need V, its maximum value. At what phase angle (Fig. 2.1) does v = V? The cosine of that angle is 1, and

$$V = \pi fD \qquad (2.4)$$

This is the "peak" or "vector" velocity, the greatest v value during any cycle.

INTERNATIONAL SYSTEM EXAMPLE

Assume f = 100 Hz
and D = 2.54 mm.
Then V = π(100)(2.54)
= 798 mm/s.

"USA" UNIT EXAMPLE

Assume f = 100 Hz
and D = 0.1 inch.
Then V = π(100)(0.1)
= 31.4 in/sec.

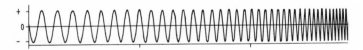

Fig. 2.2 Constant voltage e, frequency rising. In Section 2.2, e represents instantaneous displacement. Later e will represent instantaneous velocity or acceleration.

Note that if D remains constant but f rises, per Fig. 2.2, V must rise, per Eq. 2.4. If f doubles (one octave increase) V will double; such a doubling of V (dimensionally similar to voltage E) can be called a 6 decibel increase (+6 dB). This is

because

$$\text{no. of dB} = 20 \log E/E_o. \qquad (2.5)$$

If V were plotted vs. f, on log-log paper, the angle would be 45° and the slope would be +1, alternately described as +6 dB/octave.

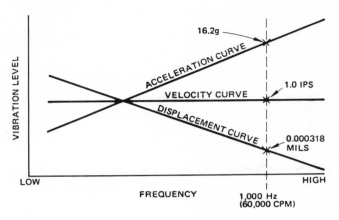

Fig. 2.3 If V were held constant, with f rising, D would drop and A would rise. See text for equations. An example is shown, for V = 1.0 in/sec at 1,000 Hz.

Conversely, if V remains constant while f rises, D must drop per Fig. 2.3. If f doubles, D must drop to 1/2 its former value; such a halving of D can be called −6 dB.

Two useful forms of Eq. 2.4 are:

$$D = \frac{V}{\pi f} \qquad (2.6)$$

$$\text{and } f = \frac{V}{\pi D} \qquad (2.7)$$

2.4 Acceleration means, to a schoolboy, rate of change of velocity. A car accelerates from 50 to 60 mph in one second; average acceleration = 10 mph/sec. Sine vibration is more complicated: a's instantaneous magnitude varies and reverses twice per cycle, per the lower graph of Fig. 2.1. When v changes slope (thus briefly zero) plot a = 0. Where v rises steepest, a = + max; where v is falls fastest, a = − max. Repeat at intermediate points, to fill in the sine wave. Note that a leads v by 90°.

Mentally or physically, ride on a child's swing. At the bottom of a stroke, v is max, but x and a are zero. a is max + when x is max −, and vice versa.

At what phase angle (Fig. 2.1) does a = max +? max −?

To get an equation for a, we differentiate or "get the slope" of Eq. 2.2, the velocity equation. The result is

$$a = \ddot{x} = -4\pi^2f^2Xsin2\pi ft \qquad (2.8)$$

In practice, nearly all instruments display D, rather than X. Since D = 2X, Eq. 2.8 becomes

$$a = \ddot{x} = -2\pi^2f^2Dsin2\pi ft \qquad (2.9)$$

We seldom use a. But we often use A, the maximum acceleration. At what phase angle (Fig. 2.1) does a = +A? −A? When the sine of the phase angle = +1,

$$a = A = 2\pi^2 f^2 D \qquad (2.10)$$

INTERNATIONAL SYSTEM EXAMPLE

Let f = 100 Hz and let D be 2.54 mm. Then A = 501 m/s².

Is this acceleration large or small? How rugged should equipment be? We lack "feel" for A expressed in m/s². So we compare 501 m/s² with the acceleration due to Earth's gravity, 9.81 m/s², by division. Now A = 51.1 grav constants or 51.1 g.

Rather than use Eq. 2.10 for calculating A, then dividing by gravitational acceleration, to express A in "g" units, we can write

$$A = 0.00202 f^2 D \qquad (2.11)$$

Occasionally you'll need a relationship between A and V at a known frequency. Simply divide Eq. 2.10 by Eq. 2.4, yielding, for A in m/s² and V in mm/s,

$$A = 2\pi f V \qquad (2.12)$$

Or, for A in g units and V in mm/s, divide Eq. 2.11 by Eq. 2.4, yielding

$$A = 0.000643 f V \qquad (2.13)$$

"USA" UNIT EXAMPLE

Let f = 100 Hz and let D = 1 inch. Then A = 19,739 in/sec².

Is this acceleration large or small? How rugged should equipment be? We lack "feel" for A expressed in in/sec². So we compare 19,739 in/sec² with the acceleration due to Earth's gravity, 386 in/sec², by division, to get A = 51.1 grav constants or A = 51.1 g.

Rather than use Eq. 2.10 for calculating A, then dividing by the gravitational acceleration, in order to express A in "g" units, we can write:

$$A = 0.0511 f^2 D. \qquad (2.14)$$

Note that if f doubles (one octave), A quadruples. A fourfold increase in A can be called a +12 dB increase, per Eq. 2.5. If A were plotted vs. f, on log-log paper, the angle would be 63.5° and the slope would be +2, alternately described as +12 dB/octave.

Occasionally you'll need a relationship between A and V at a known frequency. Simply divide Eq. 2.14 by Eq. 2.4, yielding, for A in g units and V in in/sec,

$$A = 0.0162 f V \qquad (2.15)$$

Note how Eq. 2.13 and 2.15 are shown by the rising A curve of Fig. 2.3.

2.5 Force F, we were told by Newton, equals MA. Assume M = 10 kg and the previous peak A = 501 m/s². Then the peak sine force will be 5,010 N, per the International System.

However, USA engineers modify F = MA to F = WA. Instead of 10 kg mass, they would use 22 lb. weight, and multiply that by 51.1g, for F = 1124 pounds force, very close to 5,010 N.

2.6 Piston accelerations and forces. Consider an automobile engine with 3 inch stroke (D = 3 inches or about 76 mm) operating at 6,000 RPM. What is piston V? Assume sinusoidal motion. f is 6,000/60 or 100 Hz.

INTERNATIONAL UNIT EXAMPLE

$$\begin{aligned} V &= \pi f d \\ &= \pi(100)(76) \\ &= 24{,}000 \text{ mm/s} \\ &= 24 \text{ m/s} \\ &= 86 \text{ km/hour.} \end{aligned}$$

"USA" UNIT EXAMPLE

$$\begin{aligned} V &= \pi f D \\ &= \pi(100)(3) \\ &= 944 \text{ in/sec} \\ &= 78 \text{ ft/sec} \\ &= 54 \text{ miles/hour.} \end{aligned}$$

What is the piston A? What is F, assuming piston W = 1 pound (piston M = 0.454 kg)?

INTERNATIONAL UNIT EXAMPLE

$$\begin{aligned} A &= 2\pi^2 f^2 D \\ &= 2\pi^2(100^2)(76) \\ &= 15{,}100{,}000 \text{ mm/s}^2 \text{ or } 15{,}100 \text{ m/s}^2. \end{aligned}$$

$$\begin{aligned} F &= MA \\ &= (0.454)(15{,}100) \\ &= 6{,}840 \text{ N.} \end{aligned}$$

"USA" UNIT EXAMPLE

$$\begin{aligned} A &= 0.0511 f^2 D \\ &= 0.0511(100^2)(3) \\ &= 1{,}533 g \end{aligned}$$

$$\begin{aligned} F &= WA \\ &= 1(1{,}533) \\ &= 1{,}533 \text{ pounds.} \end{aligned}$$

Why are light weight (low mass) pistons particularly valued at high RPM? Force increases with RPM². Force can be reduced by reducing piston mass.

2.7 Vibration measurements—general. Sec. 3, 4, 5 and 6 pertain respectively to the measurement of displacement, velocity, acceleration and force. Please note that subsections 2.7.1, 2.7.2 and 2.7.3 apply to measuring all four parameters.

2.7.1 Errors caused by distortion. Random vibration measurements require *true* root-mean-square (TRMS) voltmeters for indicating acceleration intensity, usually stated in RMS g units. Meters other than TRMS only "tell the truth" on pure sine waves like those on which they are factory calibrated. As such waveforms are never found on "real world" vehicles, machine tools, etc., we advise that you use TRMS meters for *all* your ac measurements.

In Eq. 2.1 we assumed pure sinusoidal motion. If that assumption is not true, then subsequent equations are not true, either.

R.M.S. VOLTS

Fig. 2.4 Voltmeter scale. Does it truly read RMS value? Courtesy Hewlett-Packard.

2.7.2 Dynamic range. dB scaling helps measure over wide dynamic ranges, as with sound or vibration intensity. Meter scales per Fig. 2.4 often appear on ac-sensing electronic meters and analyzers. Errors are often stated as a percentage of full-scale. Assume an error of ±1% of 1 volt full scale. Table 2.2 shows some readings and accuracies:

Table 2.2

Reading	Accuracy
1.0	± 1%
0.1	± 10%
0.01	±100%

To minimize errors, avoid the shaded area of Fig. 2.4.

If intensity is constant, you can take accurate readings. If intensity greatly changes, the pointer will strike the full-scale stop or will move into the shaded area. Perhaps you can change ranges, but often you cannot.

Perhaps instead your readout is Y deflection of an X-Y plotter. While your pen is high on the Y scale, readings are accurate; but when the pen moves downward, accuracy suffers. We cannot, during vibration analysis, stop the test to change Y sensitivity. (Some vibration and mechanical impedance plots require "dynamic ranges" of 10,000:1 or more.)

We need to "compress" intensity data, so that such a dynamic range can be compressed into 200 mm or 8 inches plotter Y travel, or into 100° meter pointer movement. We employ a "logarithmic converter;" its output is proportional to the logarithm of its input. Then, with appropriate scaling, we can implement Eq. 2.5.

Let's assume a 0 dB reference value of 1g RMS. Then we can construct Table 2.3 to represent other vibration intensities.

Table 2.3

Intensity RMS g	dB	Intensity, RMS g	dB	Intensity, RMS g	dB
1	0	10	20	.1	-20
2	6	100	40	.01	-40
4	12	1000	60	.001	-60

Thus a graph of vibration intensity against frequency, (as in a Tr investigation), varying from a high of 100 g to a low of 0.01 g (dynamic range 10,000:1) is compressed into 80 dB. If every Y axis inch represents 10 dB, 8 inches would represent 80 dB, or +40 to −40 dB.

Nearly every electrical handbook carries dB conversion tables to save implementing Eq. 2.5.

The international reference levels are given in Table 2.4.

Table 2.4

Sound pressure	2×10^{-5}Pa
Acceleration	10^{-5}m/s^2
Velocity	10^{-8}m/s
Displacement	10^{-11}m
Force	10^{-6}N

In Sec. 23.5, we will use decibels in connection with spectral analysis of random vibration. Then our readout will represent power. For power ratios,

$$\text{number of dB} = 10 \log P/P_o \qquad (2.16)$$

The decibel conversion calculator of Fig. 2.5 is more convenient. It now illustrates a 20 dB change. If the change is an increase, we say +20 dB. Voltage or its analog has increased by 10:1, while power has increased 10^2 or 100:1. If the change is a decrease, we say −20 dB. The voltage or its analog has dropped to 0.10 of its former value, while power has decreased to 10^{-2} or 0.01 its former value.

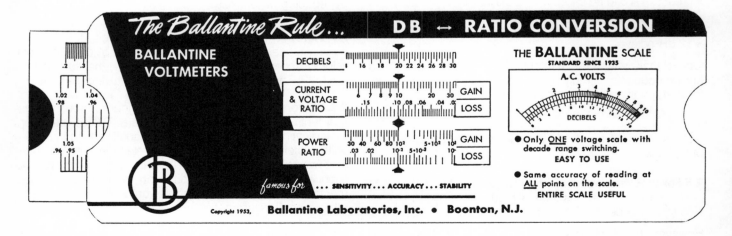

Fig. 2.5 Decibel conversion sliding calculator, more convenient than a table. Courtesy Ballantine.

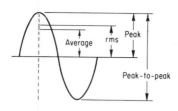

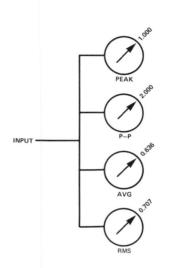

Fig. 2.6 On a PURE fixed-frequency sine wave of peak value 1.000 (or peak-to-peak value 2.000), the RMS or standard deviation will be 0.707, while the arithmetic average will be 0.636.

Fig. 2.7 Implementing Fig. 2.6, using four meters, all reading the same signal.

2.7.3 Peak vs. RMS values. Be explicit when you state values of D in inches or mm, V in in/sec or mm/s, and A in g or m/s². Further, always identify values as "maximum, peak or vector" (about same meaning for sine waves) or as RMS (Root Mean Square). Otherwise, you may be misunderstood; "10 g" could mean 41% more or 29% less to your listener. *Only* in sinusoidal vibration, per Fig. 2.6, is the RMS level 0.707 times the peak value.

2.8 Helpful vibration calculators. Cardboard calculators similar to Fig. 2.8 implement the equations of Sec. 2.2, 2.3 and 2.4. Results are not very accurate, but then vibration data are seldom very accurate. Calculations can never be correct unless signals are pure sinusoids.

Before Fig. 2.8 was taken, we placed 100 Hz at the frequency arrow. Note that a 6,000 RPM shaft would produce 6,000 cpm or 100 Hz "first order" vibration.

The lowest scales implement $V = \pi fD$ (Eq. 2.4) as well as Eq. 2.6 and 2.7. The center scales implement $A = .0511f^2D$ (Eq. 2.14), as well as Eq. 2.11 for f when A and D are known, as in setting the "crossover frequency" per Sec. 14.2. The upper scales implement $A = 0.0162fV$ (Eq. 2.15) and also solve for V when A and f are known. If you can obtain a cardboard calculator, verify it against the examples in Sec. 2.3 and 2.4.

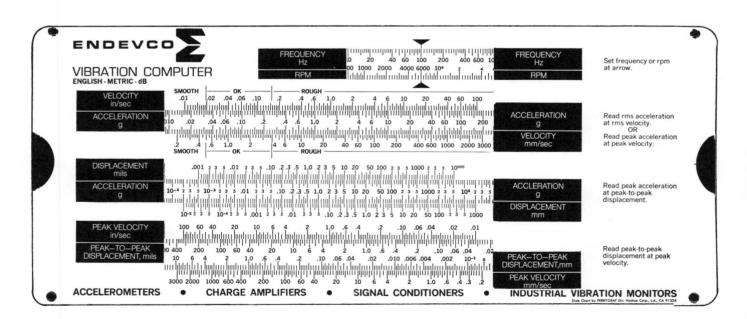

Fig. 2.8 Vibration calculator, courtesy Endevco. Many similar units offered by equipment manufacturers.

Section 3
Sensors and Systems for Measurement of Vibratory Displacement

3.0 Introduction: Why measure vibratory displacement?
Vibration measurements are important in many engineering fields.Consider electronics manufacture, with line widths 5-7 microns, soon to be lessened. Tight restrictions apply to dynamic displacements in production equipment. Here we consider some of the methods and instruments used to measure vibratory displacements. (Sec. 29.3.1 deals with torsional displacement measurements.)

3.1 Optical methods of measuring displacement have long been used.

3.1.1 Eyeball estimating of vibratory displacement is simple. We normally want the peak-to-peak displacement D in inches or in millimetres. If D is large enough, we can estimate D against a ruler. Mentally estimate the smallest D you could read with ± 10% accuracy. You might achieve this at 1/4 inch or 10 mm D.

3.1.2 Small displacements hamper optical measurements.

"USA" UNIT EXAMPLE

Let us calculate D when A = 10g (peak) at f = 1,000 Hz, using Eq. 2.12

$$A = 0.0511f^2D \text{ or}$$
$$D = \frac{A}{0.0511f^2}$$
$$= \frac{10}{0.0511(1,000)(1,000)} = 0.001957 \text{ inch}$$

(3.1)

INTERNATIONAL SYSTEM EXAMPLE

Or D when A = 10 (peak) at f = 1,000 Hz by using eq. 2.11

$$A = 0.00202f^2D \text{ or}$$
$$D = \frac{A}{0.00202f^2}$$
$$= \frac{10}{0.00202(1,000)(1,000)} = 0.00495 \text{ mm.}$$

(3.2)

With each doubling of f, D drops to 1/4 its former value, −12 dB.

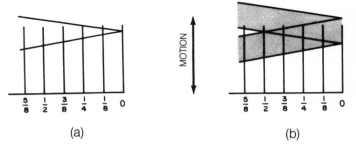

(a)   (b)

Fig. 3.1 Optical wedge or "Vee-scope," often cemented onto vibrating structure for estimating D, here 1/2 inch.

3.1.3 Optical wedge. Cement an "optical wedge" or "Vee-scope" device per Fig. 3.1(a) onto a vibrating structure. It depends upon our eyes' vision persistence above about

15Hz. Recall from Sec. 2.3 that v = 0 when x = X or −X. Your retina receives a crisp image. Between peaks (pattern is moving), your retina reports a grey image. The result *appears* to be two images, overlapping per Fig. 3.1(b). We note where the two crisp images intersect, then look directly below on the ruled scale. Here D is 0.50 inch. Accuracies of ± 5% are possible but only at larger Ds. Use this technique to measure an "unknown" D.

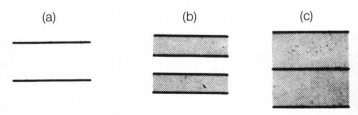

(a)   (b)   (c)

Fig. 3.2 Parallel lines inform observer when D reaches a particular value.

3.1.4 Parallel lines indicate that D has a particular value, per Fig. 3.2. Two lines are drawn 0.5 inch apart, perpendicular to the motion, per (a). For D = 0.25 inch, both grey bands are 1/4 inch high, per (b). For D = 0.5 inch, the grey bands just merge, per (c).

For calibrating at 10g (peak) and 20 Hz, carefully rule lines 0.49 inch apart, then adjust D per Fig. 3.2(c). Lines spaced 0.100, 0.200, 0.036, 0.49, etc. inch, at known frequencies, give useful Vs and As to check velocity pickups and accelerometers.

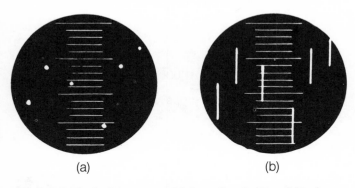

(a)   (b)

Fig. 3.3 Optical microscope aids in estimating small Ds, here 0.005 inch, with 1 mil line spacing.

3.1.5 Microscope. For measuring Ds less than 0.1 inch, use a 40X or 50X microscope (holding it stationary may be difficult) with a calibrated filar micrometer eyepiece. Focus on Scotch-lite tape, fine garnet or emery paper. With a strong side light, each target grain reflects a bright point of light, per Fig. 3.3(a). With vibration, each light point "stretches" per (b). Estimate length by comparing with the rulings. In (b), the rulings are 0.001 inch apart, and D is 0.005 inch. D values around 0.02 inch permit errors under 0.1%.

Motion pictures, particularly with camera and "strobe" light synchronized to the motion (see Sec. 16.10) help to visualize complex response modes. For "instant replay," high speed video recording seems useful.

Fig. 3.4 Holographic modal interference pattern on turbine blade at 8,240 Hz reveals tiny displacements. Courtesy Pratt & Whitney.

3.1.6 Interferometry and holography provide optical measurements of tiny displacements. With the former, 0.02 inch D can be read with errors around 0.01%. Adding a photomultiplier extends this tolerance to 10^{-5} inch D. Patterns of very small dynamic displacements can be shown under laser illumination, as in Fig. 3.4. These are laboratory (not field) techniques. See Sec. 9.6.2 for application to displacement calibration.

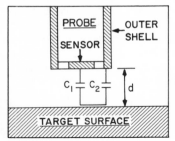

Fig. 3.5 Non-contacting capacitive relative displacement sensor. Courtesy ADE Corp.

3.1.7 Non-contacting sensors include variations in capacitance (per Fig. 3.5), inductance, eddy currents, variations in light reflected into a fiber optic probe, etc., caused by *relative* (not absolute) x variations. Optical displacement followers per Fig. 3.6 track the motion of a discontinuity on a moving

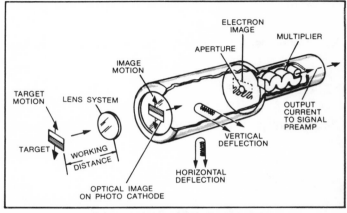

Fig. 3.6 Non-contacting optical displacement follower. Courtesy Optron.

object; the output signal represents target x. Remote readout per Sec. 3.2 is often a great benefit.

3.1.8 Hand-held contacting displacement instruments use optical and/or mechanical magnification. A probe contacts a vibrating surface, with a bright pattern (on a scale) representing D. Another unit permanently records waveforms (x vs. time) on a paper strip. Frequency is calculated from paper speed. Probe mass must not affect the motion being measured. The operator must not tremble, yet he must resist transmitted forces. As these limitations are never fully met, these instruments have little value.

3.2 Remote sensing. The main flaw is need for the operator's physical presence, sometimes difficult, often unsafe or impossible. Therefore, we convert x to an electrical signal, in order to read x remotely.

3.2.1 Attached displacement sensors convert x to an electrical signal for remote amplifiers, meters, oscilloscopes, oscillographs, tape recorders, analyzers, etc. Physical transduction methods include linkages to the sliders of variable resistances or to the magnetic rods of LVDTs (Fig. 3.7). The instrument body must be held stationary, to measure *absolute* motion. "Background" vibration often exceeds 0.005 inch D, preventing accurate measurements.

A digital optical system (U-D's DO-500) passes a light beam through two optical surfaces (one stationary, the other vibrating) each having a linear binary coded pattern of opaque and transparent lines. Relative motion modulates the light beam, and a photodetector signal produces D readouts on a counter.

3.2.2 Strain gages, bonded to structures being investigated, serve to convert x into an electrical signal. Strain readout is typically in microinches per inch of gage length. We usually place the gage where strain will be highest; a pretest, first coating the structure with a brittle lacquer, then applying vibration, helps to identify that location.

3.2.3 Displacement readout meters are essentially ac-sensing electronic voltmeters whose input signal represents x. More commonly, however, their signals come from velocity or acceleration sensors, with single or double integration per Sec. 4.4 and 4.6. Not all signals are read on meters; some are used for displacement control per Sec. 14.6.

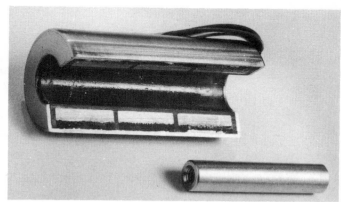

Fig. 3.7 Cutaway view of LVDT (linear variable differential transformer.) Structure being investigated drives steel rod, changing coupling between series-opposed ac-powered outer coils and center sensing coil, from which output signal is taken. Courtesy Statham.

Section 4
Sensors and Systems for Measurement of Vibratory Velocity

4.0 Introduction - Why measure vibratory velocity? Directly measuring *absolute* displacement is difficult due to "background" vibration. It is better to sense V or A, then electronically integrate into X values.

Velocity measurements have one advantage: relatively little dynamic range. Large Ds are generally associated with low frequencies, while very small Ds are associated with high frequencies; the dynamic range is 10^6. Conversely, large As are generally associated with high frequencies while only small As occur at low frequencies; the dynamic range is often 10^4. Velocity is much more uniform at various frequencies. Rather than automatically using acceleration on every program, consider velocity sensing, in spite of its drawbacks (such as sensitivity to magnetic fields).

4.1 Velocity pickups per Fig. 4.1 typically feature a seismically suspended vibrating structure. When input vibration frequency is well above sensor f_n, the magnet remains stationary. Attached to the frame, the coil sweeps through the magnetic field, inducing voltage proportional to V. Stroke is limited; the maximum D is 0.1 inch. Some are built for lower frequencies and longer strokes. Accelerations over 50g can damage them.

Velocity pickups have many advantages: they are self-contained and require no external power — no dc or ac excitation. Due to low internal impedance, they can drive long lines. They are most used from 20 to 100 Hz. Sensitivity is typically 100 millivolts (peak) per in/sec (peak) velocity. Newer units are rated -65°F to +700°F.

Velocity pickups are much larger and heavier than the accelerometers of Sec. 5. If used close to their natural frequencies, damping (usually oil or eddy-current) must be provided; this complicates non-sinusoidal motion measurements.

4.2 Velocity readout from velocity pickups is very simple. Readings on an ac-sensing electronic voltmeter can be interpreted as velocity. Or a special meter scale can indicate in in/sec or mm/s.

4.3 Calculating displacement or acceleration. If you know V and the frequency of *sinusoidal* vibration, but wish to know

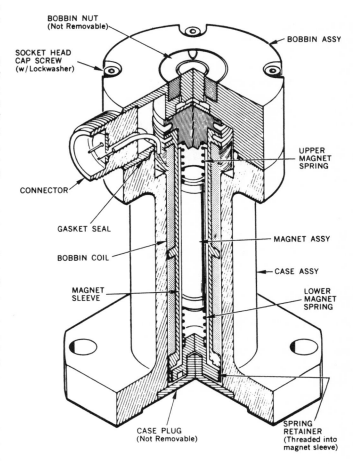

Fig. 4.1 Cutaway view of 1950's velocity sensor. Permanent magnet mass is vibration isolated from base. Courtesy CEC.

D or A, use equations 2.6 or 2.13. Or use a cardboard vibration calculator per Fig. 2.8.

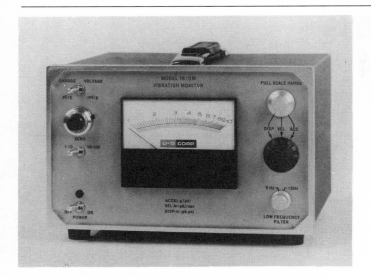

Fig. 4.2 Specialized vibration readout meter. Courtesy U-D.

4.4 Displacement readout; integration. Specialized vibration meters similar to Fig. 4.2 provide D readings in inches or mm. If velocity is sensed, an integrator per Fig. 4.3 is needed. Test this circuit with a constant-voltage test oscillator; when frequency doubles, the output voltage should drop to 1/2. R must be large enough and X_c small enough for proper integration; the "time constant" (R in ohms times C in farads) should be at least

$$\frac{10}{\text{lowest frequency at which integrator is used}}$$

and R should be at least 10 X_c.

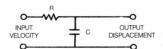

Fig. 4.3 Passive integrating network. Rolloff -6 dB/octave.

4.5 Acceleration readout; differentiation. Some meters provide a differentiating network similar to Fig. 4.4, to read A from a velocity signal. Test this circuit with a constant-voltage test oscillator; if frequency doubles, the output voltage should double. X_c must be large and R small, for proper differentiation; the time constant (R in ohms times C in farads) should be less than

$$\frac{0.1}{\text{highest frequency at which differentiator is used}}$$

and X_c should be at least 10R.

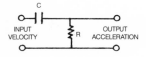

Fig. 4.4 Passive differentiating network. Slope +6 dB/octave.

4.6 Meter inaccuracies. Meters also provide for accelerometer (see Sec. 5) inputs. When that signal is integrated, the meter reads velocity; with double integration, diplacement.

Simple passive networks per Fig. 4.3 and 4.4 are limited in frequency range. Some improvement is possible by active networks. Fortunately, D is only needed from say 5 to 100 Hz; design is optimized in that region. A data is preferred above say 50 Hz; differentiators are designed for 50 to perhaps 2,000 Hz. Serious inaccuracies result outside the frequencies indicated in Table 4.1. Table 4.1 rates the meter only, not including the sensor.

Table 4.1

Function	Range	Frequency Limits	Tolerance
Velocity	0.01 to 100 in/sec	5 to 500 Hz	±2%
Displacement	0.001 to 10 inches D inches D	5 to 5,000 Hz	±2%
Acceleration	0.1 to 1,000 g	10 to 5,000 Hz	±1%
		5 to 10,000 Hz	±2%
		2 to 25,000 Hz	±5%

Note that these tolerances apply only to *pure* sinusoidal signals. If the signals from the pickup are distorted, errors can be quite large.

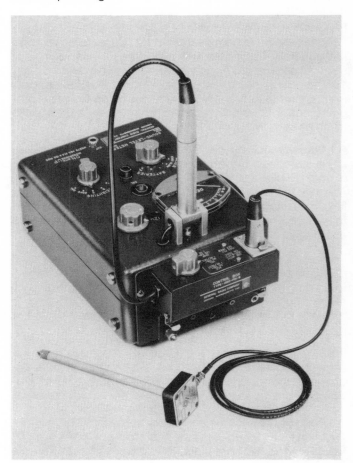

Fig. 4.5 External unit converts sound level meter to read vibration. Acceleration signals are read directly. For velocity readout they are integrated. For displacement readout they are double integrated. Courtesy GenRad.

Section 5
Sensors and Systems for Measurement of Vibratory Acceleration

5.0 Advantages of accelerometers at high frequencies. In both industrial and aerospace applications, displacement and velocity pickups (see Sec. 3 and 4) have lost favor since about 1955. Accelerometer advantages:

Some are extremely small, thus little changing vibratory responses of, say, a PCB. To determine if accelerometer mass M affects measurements, run frequency response curve A. Mount an additional inert M under your sensor. Run Curve B. If A and B greatly differ, get a lighter sensor, or use some optical or proximity device.

Some accelerometers are very rugged, routinely calibrated at 10,000g. In contrast, few velocity pickups will withstand 100g (easily damaged by rough handling).

Some accelerometer frequency ranges exceed 10,000 Hz. Most magnetic coil velocity pickups "roll off" below 10 and above 100 Hz. Those velocity sensors which combine an accelerometer, amplifier and integrator have a wider range.

Fig. 2.3 suggests constant V while frequency sweeps upward. Velocity pickup voltage would be constant. That from a displacement sensor would halve each octave per Eq. 2.6. That from an accelerometer would double each octave per Eq. 2.13. If high frequency vibration interests you, use an accelerometer. If low freqency vibration interests you, use a displacement sensor. Velocity sensing is a compromise. Sec. 29.3.3 deals with torsional acceleration.

5.1 Mechanical devices not satisfactory. See Sec. 32.1 and David A. Redhed's 1961 "A Discussion of the Problems of and the Requirements for Measuring Dynamic Acceleration with Mechanical Instruments," available from the Impact Register Co. of Champaign, IL. Mechanical-to-electrical sensors are needed for shock and vibration measurements.

A wide range of transduction principles is found in accelerometers, beyond those mentioned here. We discuss those popular in industry and aerospace.

5.2 Accelerometer flat response. Select sensors with high f_n and use them well *below* their f_n. The useful "flat" range ends at about $0.2f_n$, where sensitivity is about 4% above its low frequency value. Use a 50 kHz unit only to 10 kHz. Fig. 5.1 is based upon Eq. 1.5, the "undamped" equation. Some

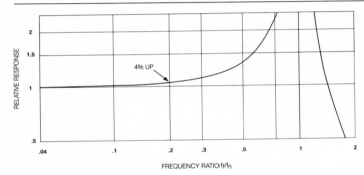

Fig. 5.1 Low-frequency portion of transmissibility graph seen earlier as Fig. 1.5.

laboratories permit $0.33f_n$, where sensitivity is up about 12% (about +1 dB). Internal filtering can extend the "flat region."

5.3 Building a crude accelerometer

might be fun. Fig. 5.2 shows a cantilever beam and strain gages R_1 and R_2. Visualize the beam horizontal, bent slightly downward by gravity forces on an imaginary mass M. Upon upward $\ddot{x}$, M's inertia further bends the beam, increasing R_1 and decreasing R_2, driving the signal (and meter deflection) negative, proportional to $\ddot{x}$. For downward $\ddot{x}$ these actions reverse.

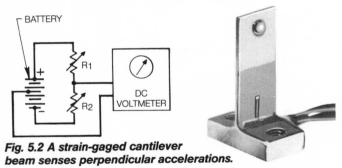

Fig. 5.2 A strain-gaged cantilever beam senses perpendicular accelerations.

For vibrations well below 1 Hz, a dc meter is acceptable. We center its pointer for zero $\ddot{x}$; it swings left for upward $\ddot{x}$ and right for downward $\ddot{x}$. We could "eyeball" +A and −A. But above say 2 Hz, our meter does not "follow" $\ddot{x}$; it just quivers about zero. We need an ac-sensing meter. Readings remain constant if A remains constant. Connect an oscilloscope to view $\ddot{x}$ waveform details.

Assume a 20 Hz accelerometer f_n. Keep A constant, and let f_f slowly approach 20 Hz ($f_f/f_n = 1$, Fig. 5.1). Note that meter readings increase. Only use this accelerometer to 4 Hz. You could damp the beam, as suggested by Fig. 1.5, to lessen resonant buildup. 10 or 15 Hz usage would be possible.

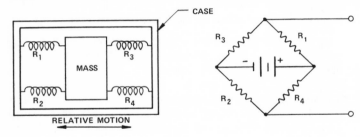

Fig. 5.3 Wire strain gage accelerometer. Note the "Wheatstone bridge" gage configuration. Courtesy Machine Design.

However, you would then have phase shift problems on non-sinusoidal vibration or shock. Better: a stiffer beam (or a smaller mass) to raise f_n, though this reduces the bending and thus our electrical response per g (our sensitivity).

5.4 A practical wire strain gage accelerometer

is suggested by Fig. 5.3. Its mass can only move along the sensitive axis. It is "sprung" by four wires which change resistance when deformed. If the case is accelerated to the right, R_3 and R_4 increase while R_1 and R_2 decrease. Output is proportional to and in phase with $\ddot{x}$. Use such accelerometers only at low frequencies. One catalog shows f_ns of 90, 155 and 250 Hz with full-scale ranges of ±1, ±2.5 and ±5g. Sensitivity must be sacrificed (e.g. heavier gage wire) for higher frequency measurements.

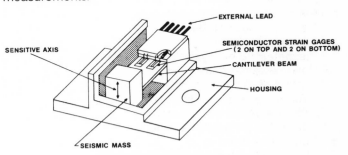

Fig. 5.4 Piezoresistive accelerometer similar in principle to Fig. 5.2 but using semiconductor strain gages. Courtesy Entran.

5.5 Piezoresistive (PR) accelerometers

per Fig. 5.4 go higher in frequency yet have good sensitivity. Semiconductor strain gages (higher gage factor than wire) sense any mass deflection due to $\ddot{x}$. A typical PR unit, with f_n = 10,000 Hz, is useful 0-2,000 Hz, range ±250g. Some are very small and light, yet sensitive. They are fine on such delicate structures as animal tissue.

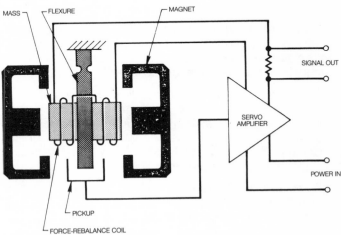

Fig. 5.5 Servo or force-rebalance type accelerometer, used as transfer standard between calibration laboratories, also used in nuclear power plants to measure earthquake severity. Courtesy Machine Design.

5.6 Servo accelerometers

are the most sensitive (can sense micro-g s of distant earthquakes) and most precise (used to navigate spacecraft). See Fig. 5.5. Case $\ddot{x}$ gives tiny internal relative motion, sensed by a displacement pickup. That signal is amplified; current passing through a coil on the mass restores mass original position and provides an output signal.

5.7 Turnover calibration. As strain-gage, PR and servo types respond to zero frequency, they are statically calibrated at ±1g by slow ±90° rotation. Imagine the accelerometer of Fig. 5.3 on its left end; gravity stretches R_3 and R_4 and shortens R_1 and R_2, giving a signal representing 1g. Rotating the unit 180° reverses the signal. These units can sense sustained unidirectional $\ddot{x}$. Calibrate them on centrifuges at 1g, 10g, etc. For most vibration work, block static $\ddot{x}$ signals with a capacitor.

electrical charge Q and voltage E when bent by any vertical $\ddot{x}$ and inertia effects on the mass. Polarity indicates direction, while signal strength indicates $\ddot{x}$ magnitude.

More practical designs give high f_ns, have wide dynamic range (fraction of 1g to thousands of g), are small and rugged. Consider the compression unit of Fig. 5.7. The crystal is heavily precompressed by a bolt. With upward $\ddot{x}$, mass inertia increases crystal compression, giving a + charge. Downward $\ddot{x}$ lessens compression, giving a − charge. Some

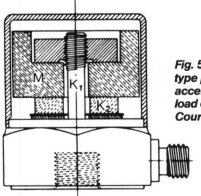

Fig. 5.6 Cantilever piezoelectric (PE) or crystal accelerometer resembling cantilever of Fig. 5.2. Courtesy Machine Design.

5.8 Piezoelectric (PE) or crystal accelerometers. Fig. 5.6 features a quartz or synthetic crystal bar to self-generate an

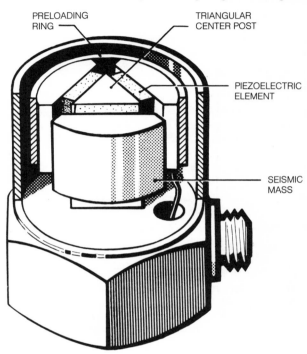

Fig. 5.9 Band secures tungsten inertia masses against shear-sensitive crystal plates, in turn against triangular column. Courtesy B&K.

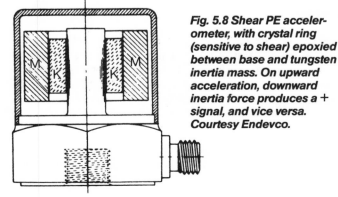

Fig. 5.7 Compression-type piezoelectric (PE) accelerometer. Note pre-load on crystal element. Courtesy Endevco.

Fig. 5.8 Shear PE acceler-ometer, with crystal ring (sensitive to shear) epoxied between base and tungsten inertia mass. On upward acceleration, downward inertia force produces a + signal, and vice versa. Courtesy Endevco.

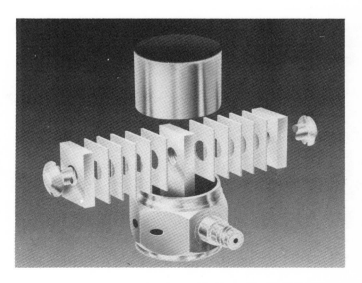

Fig. 5.10 Through bolt secures tungsten inertia masses against stacked shear-sensitive crystal plates. Courtesy Endevco.

early compression PE units were overly sensitive to attachment torque and to acoustic inputs (resembled microphones). Some showed base strain sensitivity; see Sec. 5.16.6. Some compression PE units are factory adjusted in sensitivity to 1, 10, 100 etc. pc/g (by carefully selecting inertia mass) for easy voltmeter scaling.

5.9 Voltage-sensing amplifiers. A non-electronics analogy: crystal sensors, when squeezed as in Fig. 5.12 (or bent or sheared) yield "juice" (electrons). Remove the force and the juice returns. The amount of juice (electrical charge) here measures ẍ. Electrically, this requires the "charge converter" of Sec. 5.12, for storage and measuring.

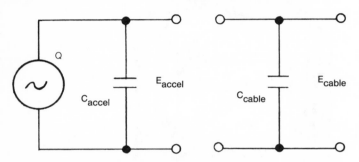

Fig. 5.11 Equivalent circuit of PE accelerometer and its cable.

For years, voltage was measured per Fig. 5.11 with voltage E proportional to charge Q divided by shunt capacitance C. Accelerometer sensitivity is given as charge (pico- or 10^{-12} coulombs) per g or as voltage per g when used with a specified shunt capacitance or with a specific cable.

$$E = \frac{Q}{C_{total}} = \frac{Q}{C_{accel} + C_{cable} + C_{ampl}} \qquad (5.1)$$

Let C_{total} be 500 pf, including 3 feet of 30 picofarads per foot cable. Let sensitivity be 10 mv per g. But with a 13 foot cable (10 extra feet or 300 more pf in C_{cable}), sensitivity will drop to

$$10 \text{ mv/g} \times \frac{500}{500 + 300} = 10 \,(0.625) = 6.25 \text{ mv/g}$$

That large drop is for only 10 additional feet. DON'T FORGET TO CORRECT FOR YOUR CABLE!! An advantage of so-called charge amplifiers (see Sec. 5.12): sensitivity does not much vary with cable length (capacitance).

5.10 Effect of discharge time constant. Measuring very low frequency vibration and long duration shocks with PE accelerometers is difficult. Electrical charge Q rather quickly "leaks off" through the 10^7 ohm input resistance R of a typical readout instrument (meter, analyzer, 'scope or tape recorder, etc.) terminating our cable. If our vibration is at say 1,000 Hz, we can read its magnitude on an ac-sensing electronic voltmeter; successive cycles of signal arrive before voltage leaks off.

The discharge time constant T of our accelerometer + cable + instrument equals R × total C; that is, T = RC. Consider a situation where R is 10^7 ohms and C is 10^{-9} farad, or 1,000 pf. Then T = RC = 0.01 second.

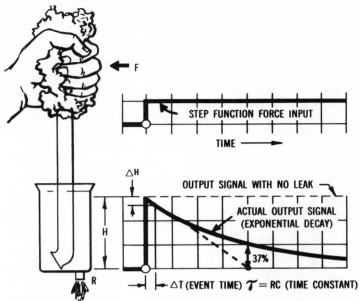

Fig. 5.12 Hydraulic-to-electric analogy: exponential decay of depth H (voltage) is due to fluid (electrical) leakage. Courtesy Kistler.

Fig. 5.12 (upper) shows a "step function" of ẍ. If T were infinitely long, the signal would exactly represent ẍ. With finite T, the signal drops exponentially. At its original slope, the signal would reach zero in T = 0.01 second.

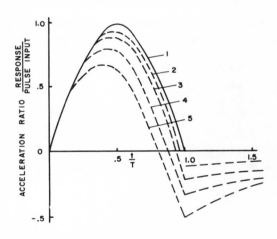

Fig. 5.13 Effect of measuring half-sine shock pulse with insufficient RC time constant. RC should be at least 20 times such a pulse duration.

This works well for measuring very short shocks, duration say 1 μsecond. But when measuring longer shocks, say 50 μs, we are in trouble. This works well for measuring vibrations above 1/0.01 sec or 100 Hz (error about -1.5% at 100 Hz.) Fig. 5.14 shows severe rolloff below 100 Hz. At 50 Hz we'll read 5% low; at 9 Hz, 50% low. We must correct our data, a nuisance.

We could add a large shunt capacitance C. However, we would lose voltage sensitivity, since E = Q/C. Small ẍ would be hard to measure.

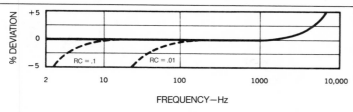

Fig. 5.14 Low-frequency response can be quite flat with charge-sensing electronics. Voltage-sensing systems are plagued with low-frequency rolloff unless RC time constant is large.

5.11 Voltage signal conditioning. The 1950s solution was voltage-sensing preamplifiers (cathode followers — later, emitter followers), having high input R, typically 10^8 ohms. Now, with C = 1,000 pf, T = RC = 0.1 sec. We can accurately measure longer shocks and vibration down to 10 Hz, per Fig. 5.14. At 5 Hz, response will be 5% low. For many applications, this is fine. Sec. 5.10 suggests that PE units are poor for measuring long shocks or very low-frequency vibration. We need even higher input R, but cable or connector leakage due to skin oils, salt water immersion or salt spray, moisture, etc., can (unknown to test personnel) lower R and seriously degrade long duration shock and/or low frequency vibration measurements.

Some preamplifiers are built into readout instruments; often better are "line driver" units, receiving PE signals over a short cable, then sending low impedance signals (less susceptible to noise) to relatively remote downstream instruments.

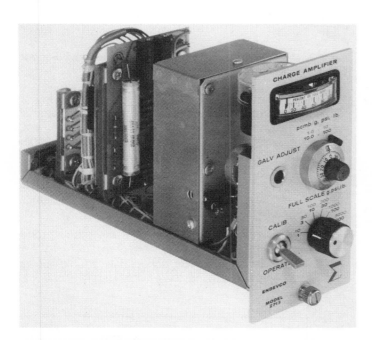

Fig. 5.15 Rack-mountable "charge amplifier" for use with PE sensors. Courtesy Endevco.

5.12 Charge signal conditioning avoids these low-frequency errors. Moderate shunt capacitance and moderate contamination of cables and connectors have little or no effect. Response can go to extremely low frequencies per

Fig. 5.16 Charge to voltage converter, to precede further amplification in readout instruments. Courtesy Metrix.

the solid line of Fig. 5.14. (Most PE systems stop at 0.2 or 0.5 Hz per Sec. 5.16. One firm offers 10^{12} ohm input resistance with T to 1 sec, and low frequency measurements to 0.02 Hz. Regardless, PRs seem easier to use.)

Fig. 5.17 Charge to voltage converter to be located close to PE sensor, then driving a low impedance relatively long transmission line to other instruments. Courtesy Endevco.

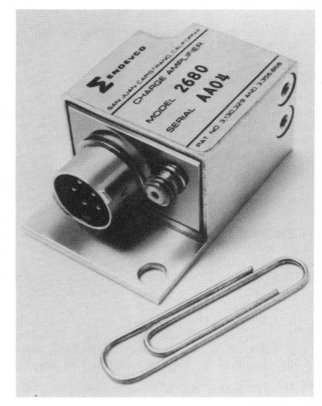

Fig. 5.18 Rugged charge converter and amplifier to condition PE signal for telemetry system. Courtesy Endevco.

The name "charge amplifier" is a misnomer. Charge is converted to an analog voltage (*representing* charge), then amplified. For the circuit details of a specific unit, see the manufacturer's data. Special "charge converters" permit low (i.e. 10k ohms) PE source resistances when PE accelerometers operate at say +1,400°F.

Measuring intense but short shock pulses can be difficult for signal conditioners and other parts of the instrumentation chain. See Sec. 32.2.4 on "pyro" shock measurements.

Fig. 5.20 Multichannel programmable signal conditioning and control system. Courtesy Endevco.

Secretarial chores are greatly eased by auto-ranging analog amplifiers or by the digital unit of Fig. 5.20. It guides the operator through his setup routine. Controlled amplifiers (nearby or distant via a modem) are keyboard programmed, then settings are stored for repeat testing. Bandpass filtering can be inserted. All channels are automatically calibrated with data output in engineering units. Signal magnitudes are scanned during a test; if limits are exceeded, the operator is alerted and gain is changed automatically. Initial gain settings and all gain changes are logged for fully documenting a test.

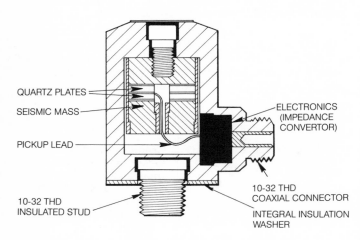

Fig. 5.19 Solid-state amplifier inside sensor has many advantages. See text. Courtesy Kistler.

5.13 Amplifier built into sensor. Units containing solid-state amplification, as in Fig. 5.19, protect the signal against contamination, leakage (changes in R) and changes in shunt capacitance. High electrical output (several volts per g possible) and low (say 100 ohms) output impedance minimize electrical interference, even with inexpensive, rugged wire. Some cost increase is partly offset by simpler external electronics. Ambient temperature may be limited to +250°F. Range changing is hampered, so extremely large (or small) $\ddot{x}$ measurements are difficult.

5.14 Electrical filtering. Measurements are plagued with "noise," unwanted electrical signals. How to reduce that noise, improve our "signal/noise ratio"? Restricting the bandwidth over which we measure is helpful. Example #1: accelerometer mechanical "ringing" at say 50kHz may mask a structural resonance. Insert a low-pass filter, attenuating all signals above say 12 kHz to ease data interpretation. Example #2: long-duration "noise" signals can result from thermal transients (see Sec. 5.16.2). Insert a high-pass filter, attenuating all signals below say 1 Hz to aid data interpretation.

5.15 Programmable signal conditioners. Order all your sensors at the same sensitivity, interchangeable. Otherwise, test personnel twiddle knobs, set switches and take many notes. Errors are made, especially when a test requires say 50 accelerometers plus various other pickups.

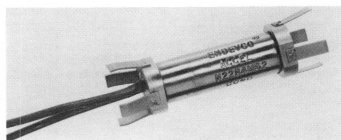

Fig. 5.21 Dual-axis PE accelerometer for nuclear environment. Courtesy Endevco.

5.16 Accelerometer system difficulties, or "What the salesman didn't tell me." Design of transducers attempts to maximize sensitivity. And to minimize the effect of other environments (such as temperature, acoustical fields, pressure fields, electrostatic and electromagnetic fields) on the signal. Consider the sensor of Fig. 5.21, which goes into a nuclear fuel rod. Hopefully the sensor materials will "ignore" gamma and neutron flux.

Recent calibration (sensor alone or entire system) *under laboratory conditions* is no guarantee of reliable field or test lab data. Users must not blindly accept data, should question the instruments, the cabling, the environmental effects, etc.

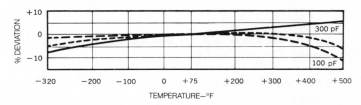

Fig. 5.22 Ambient temperature effects on a PE accelerometer's sensitivity.

5.16.1 Ambient temperature effects. Fig. 5.22 shows voltage sensitivity varying with temperature, also improvements with added shunt capacitance. Curves often differ widely between voltage and charge response; use charge- or voltage-sensing electronics, depending upon which curve (charge or voltage sensitivity) is flatter.

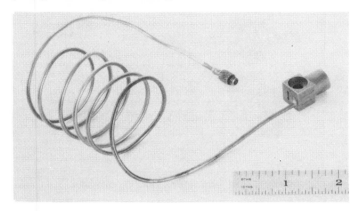

Fig. 5.23 PE accelerometer useable to 1,400 F. Courtesy Endevco.

A brief high temperature event might not be fatal, but check calibration to be sure. (If the *crystal* reaches its Curie temperature, goodbye!) Some PE materials change resistance and/or capacitance (and thus low frequency response) at high temperatures.

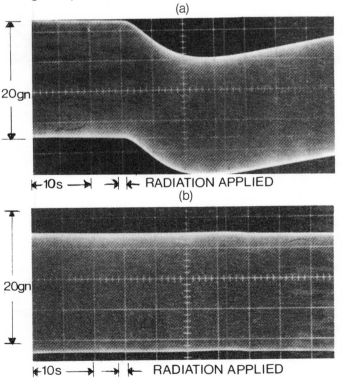

(a)

20gn

←10s→ →|← RADIATION APPLIED

(b)

20gn

←10s→ →|← RADIATION APPLIED

Fig. 5.24 (a) PE accelerometer zero shift due to one second thermal transient (continually experiencing 20g peak at 100 Hz); readout via charge amplifier into 'scope. (b) Same experiment, but here the signal is high-pass filtered (down 5% at 2 Hz).

5.16.2 Transient temperature effects. Fast temperature changes generate a "pyroelectric" charge in certain PE materials. With good design, crystals are mechanically protected from housing expansion. You can slow temperature changes by circulating air or water, by using a heat sink or radiator, or by thermal insulation. When the temperature changes gradually, the resulting signals are extremely low in frequency; use a high-pass electrical filter to block signals below say 1 Hz. However, "temperature shocks" can overload or even damage your amplifier. ANSI S2.11-1969 procedure involves plunging a (previously stabilized at 82°F) accelerometer and mounting block into ice water. Signals representing 0.2g to 51g have been reported. Don't permit PE units to be open-circuited while temperatures change.

When selecting a PE sensor for use near explosives, or any IR source, be sure it does not surge as in Fig. 5.24. See NBS Technical Note 855, by Vezzetti and Lederer. They flashed radiation 19 cm. from a 600-watt incandescent lamp onto PE accelerometers and saw zero shifts as high as 640g. They noted that quartz PEs seemed least affected. Non-uniform heating of the sensor was felt probable. Pyroelectricity? Differential thermal expansion? They recommend a high-pass filter to block zero shifts from readout instruments.

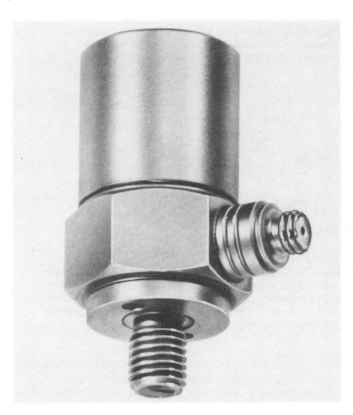

Fig. 5.25 Classical accelerometer mounting technique: a shouldered stud. Courtesy Kistler.

5.16.3 Accelerometer mounting techniques. Fig. 5.25 shows the classical way (always used in standards laboratories) to attach an accelerometer. Tighten it with a torque wrench adapter similar to Fig. 5.26(A). Attain a tensile preload greater than F = MA to prevent "chattering" between sensor and structure. Drill the mounting hole perpendicular to the surface. Thread it deeply so the stud cannot "bottom." The surface should be flat and smooth. Above 5,000 Hz, oil or silicone grease aids the mechanical connection.

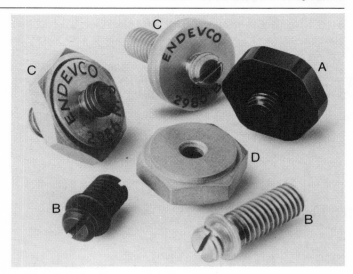

Fig. 5.28 Several accelerometer mounting devices. Courtesy Endevco.

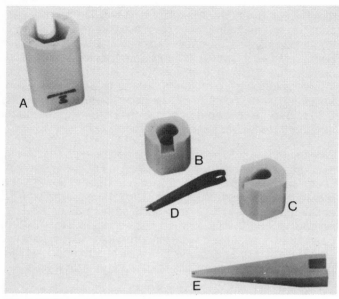

Fig. 5.26 Wrenches and adaptors for use with accelerometers. Never strike an accelerometer to break cement bond. Twist it off. Courtesy Endevco.

STRIKE AN ACCELEROMETER! Some laboratories mount accelerometers by a thin film of wax. Ask PCB for their report on Petro-Wax. Avoid double-stick tape.

Fig. 5.27 A rugged "industrial" housed PE accelerometer. Note stainless steel casing, sturdy ARINC base, replaceable armored cable and screwed-on connector. Available "diode barrier" prevents possible sparking and permits inflammable usage. Courtesy B&K.

Fig. 5.29 An aluminum block, shaped to match missile contour, for orthogonal mounting of two accelerometers.

Can't drill a mounting hole? Cement an attachment stud per Fig. 5.28(A). No flat location? Machine a bracket or better a block (avoid plastics except at very low frequencies) to match the structure's contours (see Fig. 5.29.) But watch out for resonances. Dynamically evaluate brackets and blocks as with test fixtures (see Sec. 19.13).

No screw connection on your accelerometer? Cement it directly. Clean the base (no sanding!!) and the location with a suitable solvent. Use a very thin cyanoacrylate or dental cement film. Thoroughly evaluate results before adopting a new material. See Fig. 5.30. After the test, use one of the wrenches of Fig. 5.26 to twist off the accelerometer. NEVER

Machine screw mounting per Fig. 5.28(B) is best dynamically. But don't use calibration data taken with a machine screw when mounting another way. The insulated studs of Fig. 5.28(C) help prevent "ground loops" (Sec. 5.16.5). These introduce some compliance, but are satisfactory to about 2,000 Hz.

In-service vibration is seldom uniaxial. How to measure multi-axis motion? Use a "triaxial" assembly as in Fig. 5.32 to sense vibration along three major axes.

Special purpose mountings are sometimes needed. See, for example, the seat mounting pad of Fig. 0.7.

EFFECT OF "ROCKING" DUE TO FAULTY CEMENT BONDS ON ACCELEROMETER RESPONSE

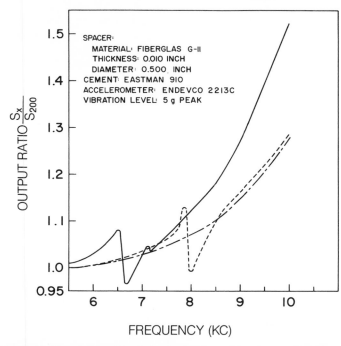

SPACER:
  MATERIAL: FIBERGLAS G-II
  THICKNESS: 0.010 INCH
  DIAMETER: 0.500 INCH
CEMENT: EASTMAN 910
ACCELEROMETER: ENDEVCO 2213C
VIBRATION LEVEL: 5 g PEAK

Fig. 5.30 Effect of "rocking" due to faulty cement bonds. Smooth resonant buildup is proper. "Glitches" indicate faulty bonding. After Mangolds.

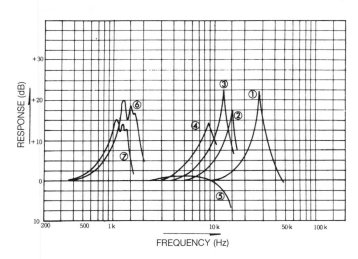

Fig. 5.31 Experimental results of several mounting techniques: (1) Tightly screwed. (2) Fixed with insulating adapter. (3) Fixed with hexagonal flat attachment and instantaneous bonding agent. (4) Fixed with attraction magnet. (5) Fixed with hexagonal flat attachment and paper adhesive tape. (6) Held by round bar attachment. (7) Held by stylus bar attachment. Courtesy Rion.

The hand-held probe of Fig. 5.33 speeds multi-axis investigations, as during a sine resonant search. Don't overly trust the results, however; the probe itself may resonate, limiting the useful frequency range.

On flexible test items such as PCBs, watch out for modal changes due to accelerometer mass, also for local stiffening. Use the smallest available sensors, that cover least area.

Fig. 5.32 Triaxial array of three accelerometers. Courtesy Kistler.

5.16.4 Cable noise, especially between a high impedance or charge mode PE sensor and first amplifier stage, is often troublesome. The cable must transmit a "high fidelity" signal. Shielding minimizes electrical pickup. Cable disconnects at the sensors can be noisy; integral cables or soldered connections are quietest. Cable capacitive loading of the charge converter input can degrade resolution and signal.

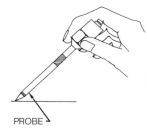

Fig. 5.33 Hand-held probe for quickly exploring response modes.

PROBE

After much flexing, cables can internally generate low frequency electrical signals or "noise," yet electrically check OK. See Fig. 5.34. When flexed, the shield separates briefly from the dielectric, generating a *triboelectric* charge. Excess electrons pass onto and then along the center conductor to the amplifier input. The readout instrument can't tell cable flexing from sensor motion. Check cables per ISA's RP 37.2 Para 5.3.2. More recent cables feature conductive coatings on the dielectric. Triboelectric charges distribute themselves locally, and less noise reaches the amplifier input.

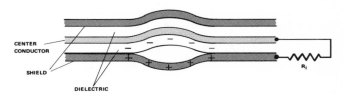

CENTER CONDUCTOR

SHIELD

DIELECTRIC

R_i

Fig. 5.34 Cross section of accelerometer cable construction. At center, a stranded copper wire, then a Teflon layer, braided soft copper and finally plastic jacket (not shown). Bond separation gives cable noise.

Watch out for large-D low frequency resonant cable motions, especially troublesome with a "strain-relief loop" near the accelerometer, per "C" of Fig. 5.35. Resonance can cause internal separation. Long shock pulses can damage cables.

Fig. 5.35 Several possible ways to secure accelerometer cable. C is worst. After Mangolds.

Only one experimenter has written on securing cables (B. Mangolds, SHOCK & VIBRATION BULLETIN 33, Part 3). He recommends not restraining laboratory PE cables. (Restrain them outdoors or on a shock test carriage.) His method: run cables in rigid conduit to a junction box directly over your test. Coil and secure excess cable to the box. Drop the needed length down to the accelerometer (no restraint). You will receive contrary advice on this topic, urging taping to nearby structure per "D" of Fig. 5.35. If your experiments show that such restraint reduces cable noise, please communicate with TIT. And publish your data. Meantime, we promote Mangold's method, used in leading laboratories, worldwide.

Cables should be light and flexible, so cable resonance and whipping little distort the case (might stress the crystal). Few technicians can effectively repair cables. Yes, cables are expensive. But possible loss of valued data outweighs cable cost. When in doubt concerning a cable, throw it away!

In humid environments, place a drip loop close to the sensor. Clean the cable and fitting with acetone, then seal with Dow Corning 732 RTV.

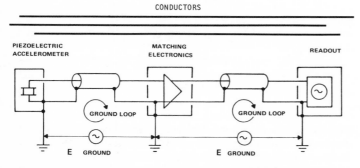

Fig. 5.36 "Ground loops" occur when signal paths are grounded at more than one point, when a voltage (due to current in other conductors or through Earth) exists between those points.

Accelerometer cables flexing at high velocities generate heat. Use high temperature cable or provide cooling.

5.16.5 Ground loops per Fig. 5.36, can result in "noise" masking the signal of interest. Only ground one point: usually your readout device input. Electrically insulate signal amplifiers and sensors. Simply ungrounding the sensor or using an internally ungrounded sensor may not be enough. An internal Faraday shielded sensor may be needed, along with double-shielded cable. Try a differential amplifier. See Ralph Morrison's 1967 text "Grounding and Shielding Techniques in Instrumentation," (Wiley). Also TAR-109, "Noise Immunity," from Vibra-Metrics, Inc. at 385 Putnam Ave., Hamden, CT 06517.

Consider fiber optic cables to interconnect various units, particularly when transmitting digital signals. This should eliminate ground loop problems.

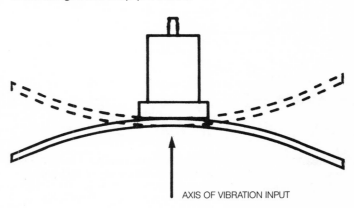

AXIS OF VIBRATION INPUT

Fig. 5.37 If there's no translation, there should be no signal.

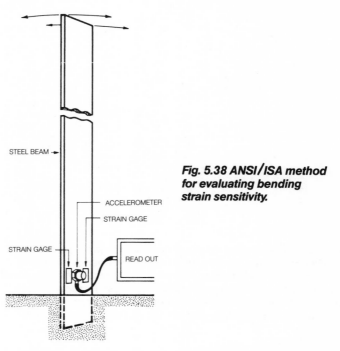

STEEL BEAM

ACCELEROMETER

STRAIN GAGE

STRAIN GAGE

READ OUT

Fig. 5.38 ANSI/ISA method for evaluating bending strain sensitivity.

5.16.6 Base bending strain caused some early compression PE units to generate large unwanted signals. Test structure flexing per Fig. 5.37 could reach the crystal. Follow ANSI S2.11-1969 Para. 8.5 or ISA Standard RP37.2 with known dynamic strains to evaluate accelerometers; see Fig. 5.38. Select sensors with low strain sensitivity. Alternately, mount on a stud (Fig. 5.28C) to transmit motion but not bending.

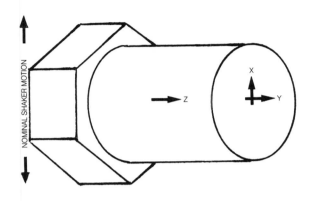

Fig. 5.39 Identification of accelerometer axes for discussion of lateral sensitivity.

5.16.7 Lateral sensitivity refers to motion perpendicular to sensor axis. Consider the accelerometer of Fig. 5.39 (one of three in triaxial array per Fig. 5.32.) Ideally, it only senses z motion. If any signal results (as during shaker evaluation), we might wrongly assume z shaker motion when shaker motion is all x directed. An accelerometer with high lateral sensitivity can make a good shaker look bad.

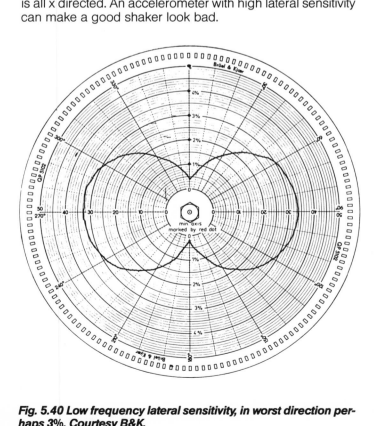

Fig. 5.40 Low frequency lateral sensitivity, in worst direction perhaps 3%. Courtesy B&K.

Manufacturers typically claim that lateral sensitivity is less than 3% of axial sensitivity. But at what frequency? Most only check in the range 12 to 30 Hz, with results per Fig. 5.40. B&K warns that at high f_S lateral sensitivity can be significantly higher, due to a lateral resonance. See Fig. 5.41, which may confirm unpublished 1963 work at the Navy's Metrology Engineering Center, Pomona, CA. High frequency lateral sensitivities were rumored to approach 100% on advertised 3% (low frequency) units.

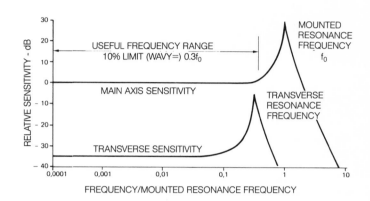

Fig. 5.41 Idealized graphs of main axis and lateral sensitivities over wide frequency range. Lateral sensitivity is small at low test frequencies, but a small transverse $\ddot{x}$ at the transverse f_n could be misinterpreted as large axial motion. Courtesy B&K.

Most new accelerometers have very low lateral sensitivity. But when a sensor is dropped or otherwise severely shocked, it may suffer damage. Repeat important investigations (which suggest excessive lateral motion) with new sensors to insure that motion is accurately reported.

5.16.8 Amplifier overloads are discussed in Sec. 32.2.4 on "pyro" shock. Undetected overloads, highly insidious causes of faulty measurements, can occur along the signal path between sensor and analyzer. "Clipping" drastically changes the signal to an analyzer. The resulting spectrum contains significant errors (generally at higher frequencies).

Never exceed the input and output voltage limitations of any amplifier along your measurement chain. Using known value sine test inputs, look at signals (with your 'scope) at various checkpoints, particularly before any filtering. Advanced power supplies (furnishing power to amplifiers inside sensors per Sec. 5.13, feature "overload" indicators.

5.17 Verifying behavior. Formal calibration per Sec. 9 is rightly performed by specialists. However, considering the many error sources discussed above, it is prudent to check sensors, systems and operator skills. Examine mounting effects on system behavior. With digital 'scopes and analyzers, verifying accelerometer behavior is relatively easy; gravimetric methods can extend "turnover" calibration per Sec. 5.7 to greater intensities.

Section 6
Sensors and Systems for Measurement of Vibratory Force

6.0 **Why measure vibratory force?**
6.1 **Force sensors and signal conditioning**
6.2 **Driving point sensors**
6.3 **Useful ratios**
6.4 **Readout instruments**
6.5 **Calibration**
6.6 **Why measure mobility?**
6.7 **Controlling a vibration test**

6.0 Why measure vibratory force? You probably routinely measure motion, using displacement or velocity sensors or, most commonly, accelerometers. Why would anyone measure F? "Aren't F and A practically the same thing, related per F = MA?" Not in dynamics.

Fig. 6.1 An EM shaker driving a structure through a driving point sensor (impedance head) which measures force applied to linkage driving test structure; sensor can also measure acceleration, though here a separate accelerometer is being used. Courtesy Land-Air.

Internally, accelerometers employ force measuring sensors. Refer to the compression PE accelerometer of Fig. 5.7. Visualize the crystal being located instead in the structural path between a shaker and some load (rather than driving the sensor's internal tungsten mass); now it is a PE force sensor.

When discussing accelerometers, we often state that the signal represents $\ddot{x}$. This works out in practice because the tungsten mass *behaves* like a mass (a rigid body) at all frequencies of interest.

If we wish to learn the vibratory characteristics of "real world" structures, we must measure force *and* motion. Have you never encountered behavior that didn't "fit" the classical F = MA? Where your shaker, for example, developed more (or less) A than its F rating predicted? Where the load was surprisingly easy (or hard) to drive? Certainly the explanation was "resonance," but to fully understand its behavior you require (1) how strongly the load responded and (2) the effective force you applied. Unfortunately, many test personnel ignore (2). A term used by structural dynamicists: "apparent mass;" if mass didn't change (some engineers vehemently deny it *can* change), that term wouldn't exist. Experimentalists relate force and motion (stimulus and response).

Even in 1984 a few test engineers (and their customers) fail to recognize need for force measurements and data. But we expect an increase in awareness. Torsional force and impedance measurements are discussed in Section 29.2.4.

6.1 Force sensors and signal conditioning. How does a force sensor work? Just like a "load cell," per Fig. 6.2. But in dynamics work PE force sensors per Fig. 6.3 are better, since their frequency range is typically 5 - 5,000 Hz. Visualize the crystal experiencing a driving force on one surface (possibly an aircraft structural point), and passing that force onward to some "black box." (Alongside the black box, an accelerometer could measure structural motion. Or, per Fig. 6.4, the force sensor might aid in modal testing per Sec. 1.4. Later, in Sec. 6.7, we'll discuss how both instruments aid in controlling a vibration test.)

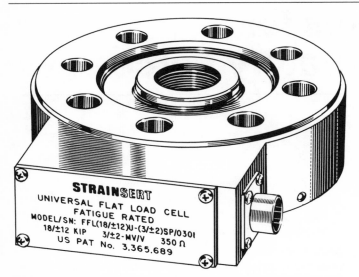

Fig. 6.2 A strain gage force sensor or "load cell;" useful in, for example, static determination of weight. Courtesy Strainsert.

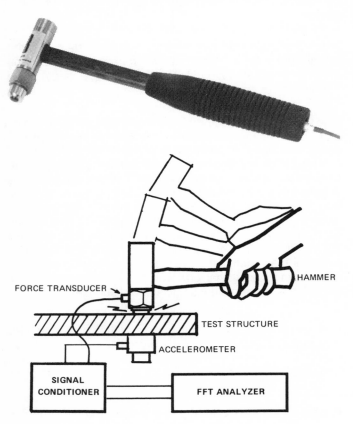

Fig. 6.4 Hammer used to impulsively excite a structure. Transducer converts applied force to electrical signal. Response accelerometers sense structure's motion. Signals are processed for modal analysis. Courtesy PCB.

Signal conditioning for PE force sensors is identical with signal conditioning for PE accelerometers; see Sec 5.11, 5.12.

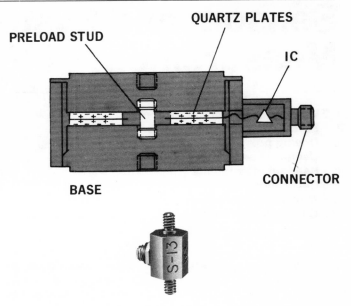

Fig. 6.3 PE force sensors, cutaway and external views. Courtesy PCB and Wilcoxon.

6.2 Driving point sensors (or impedance heads - see Figs. 6.1 and 6.5) are essentially a combined (within one housing) force sensor + accelerometer. They vary in size, per Fig. 6.6.

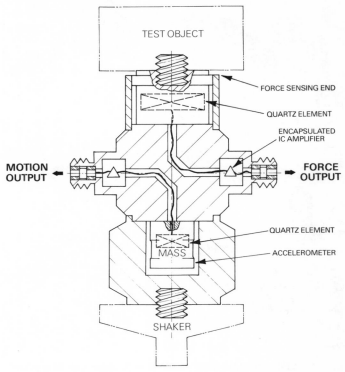

Fig. 6.5 Cutaway view of driving point sensor (impedance head), which generates two signals: (1) motion at driving point and (2) applied force. Courtesy PCB.

Fig. 6.6 Three sizes of driving point sensors. Courtesy Wilcoxon.

6.3 Useful ratios. Structural dynamicists make use of six ratios, three inverse pairs, per Table 6.1.

Receptance and admittance are alternate terms for compliance. Impedance and mobility are most common. Velocity signals are obtained by integrating an accelerometer signal after charge to voltage conversion. Displacement signals are obtained by two integrations.

6.4 Readout instruments, cabling, charge to voltage conversion, amplification and/or magnetic storage used with force sensors closely resemble those used with accelerometers. Meters are scaled in newtons or pounds (peak for sine vibration and for shock, and RMS for random vibration).

6.5 Calibration of a force sensor requires driving it with an EM shaker per Sec. 9.9. Let the gage drive an appropriate mass (must behave as a mass at all f_fs). Measure and record the force signal (over a range of f_fs) while maintaining say 10g at the shaker. Repeat with "zero load" attached. (Actually, a small load, part of the gage, is driven by the crystal.) Then, at each f_f, subtract the second signal from the first. (Mass compensation can be important at test item resonance, where dynamic mass is small.) Finally, divide the net force signal by the mass used and again by the acceleration, say 10g. This will give you "electrical units" divided by "mechanical units" per Sec. 9.0. Alternately, drive through your force gage into a calibration standard force gage (for comparison calibration similar to Sec. 9.7) into a mass.

6.6 Why measure mobility or any of the ratios per Table 6.1? You may need to measure a structure's dynamic properties (mass, stiffness, damping), possibly to verify a mathematical model. You may need to identify f_ns and mode shapes; perhaps you will perform modal tests per Sec. 1.4.

6.7 Controlling a vibration test by means of a force sensor (in lieu of or in addition to, an accelerometer) seems strange to "old timers." Traditional tests have been controlled by accelerometers; see Sec. 14.6 and 23.7.

Visualize an aircraft flight test (of an actual unit or a dynamically similar dummy). The support structure's driving force (into the unit) and the motion at their interface are both being recorded. Unfortunately, in most past flight testing, only motion has been sensed and recorded. Force has been ignored. Can you appreciate how measuring only the motion would measure only part of the unit's dynamic environment? How its internal parts might behave differently if it were attached to a different location or to a different aircraft? That our usual sensing/specifying/controlling only shaker motion, without accounting for the *mobility* of the test fixture and shaker, might be incomplete? How we might be overtesting at some f_fs and undertesting at others?

Controlling sine or random tests by a force sensor is quite simple. Just substitute that signal in lieu of your accelerometer signal. The task of the sine servo or the random equalizer is actually simplified, because *force* is being controlled, rather than motion; EM shakers are basically constant-force (for a constant current) devices.

Table 6.1

INERTANCE or ACCELERANCE	Acceleration ÷ Force
DYNAMIC MASS	Force ÷ Acceleration
MOBILITY	Velocity ÷ Force
MECHANICAL IMPEDANCE	Force ÷ Velocity
COMPLIANCE	Displacement ÷ Force
DYNAMIC STIFFNESS	Force ÷ Displacement

Section 7
Introduction to the Analysis of Complex Vibration

7.0 Introduction - the sine wave. We have almost exclusively considered sinusoidal vibration. It exists only at one frequency; no other measurable vibration is present. When sensed and displayed on a 'scope, we get the "clean" pattern of Fig. 7.1. Note that this is the only signal which most meters read accurately. Most meters err by large factors on other waveforms.

7.1 Sine vibration seldom exists outside the calibration lab. The vibration of actual structures, engines, machine tools, appliances, vehicles, etc. is almost never concentrated at a single frequency. Rather, the vibration of such equipments is often *complex*, concentrated at two or (usually) more discrete frequencies.

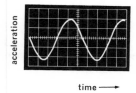

Fig. 7.1 Oscilloscope (time history or time domain) and spectral (frequency domain) displays of a sine wave. All energy is concentrated at a single frequency f.

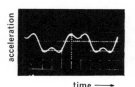

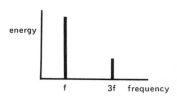

Fig. 7.2 Oscilloscope (time history or time domain) and spectral (frequency domain) displays of a complex wave. All energy is concentrated at two discrete frequencies: f and 3f.

7.2 Examples of complex vibration. The time history and spectrum of Fig. 7.2 show vibration concentrated at two discrete frequencies that happen to have the ratio 3:1. You could describe that vibration as 70% fundamental vibration at (for example) 100 Hz + 30% third harmonic vibration at 300 Hz. What might cause such a vibration? In sinusoidal vibration testing we often see third harmonic content in the power amplifier output (see Sec. 13.6) coinciding with a resonance in shaker, fixture or test item. The spectral display of Fig. 7.2 quantizes the vibration existing simultaneously at the two frequencies. If the test item fails, which frequency did the damage?

A rotating machinery difficulty: an engine turning at 6,000 RPM with some rotational unbalance generates a fundamental at 100 Hz; an accessory driven by a 3:1 speed-increasing gear system, also rotationally unbalanced, generates a 300 Hz harmonic. (Gear ratios of exactly 3:1 are rare, but this example simplifies our explanation).

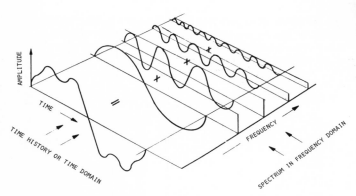

Fig. 7.3 Relation between time and frequency domains for a complex wave having its fundamental at 10 Hz and harmonics at 30, 50 and 90 Hz.

Consider the isometric view of Fig. 7.3. If we look from the left, we're in the "time domain." We're viewing 'scope pictures of successively, a complex wave, then its consituent signals. First, the fundamental with one cycle in say 0.1 second. Then the third, fifth and ninth harmonics, with 3, 5 and 9 cycles in 0.1 second. The four frequencies must be 10, 30, 50 and 90 Hz. If we view from the right, we're using a spectrum analyzer, and we're in the frequency domain. We see signal peaks at 10, 30, 50 and 90 Hz. We can measure the individual signal magnitudes.

7.3 Mechanical vibration analyzers function much like the electronic analyzers of Sec. 7.9 and 7.10. Visualize an adjust-able-length cantilever beam. Hold it against an "unknown" vibration and adjust its length until the beam resonates. Then consult markings on the beam (previously calibrated on a precision shaker) to read the "unknown" frequency. With most vibrations, several "unknown" frequencies can thus be identified. Tunable electronic analyzers functionally resemble this unit.

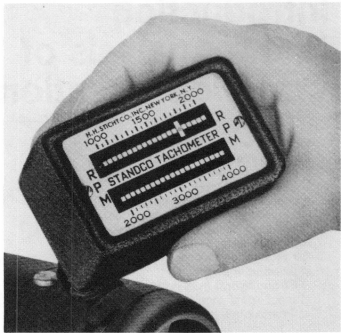

Fig. 7.4 Handheld multi-reed mechanical vibration analyzer. Motor's slight shaft unbalance creates first-order vibration at 1,750 RPM. Courtesy H.H. Sticht.

The multi-reed analyzer of Fig. 7.4 features numerous reeds which can respond simultaneously to fs that may be present. With many reeds, fairly accurate spectrum analysis is possible. The multiband analyzers of Sec. 7.9 functionally resemble this unit.

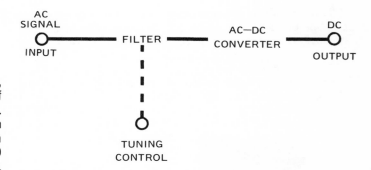

Fig. 7.5 Concept of (mechanical but usually electronically) tuned filter for obtaining details of spectrum.

7.4 Tuneable analog analyzers are suggested by Fig. 7.5. They can identify frequencies f_1 and f_2 and their magnitudes, Fig. 7.2. Analyzers are used to study various physical phenomena (vibration, sound, pressure, force, etc.) converted into electrical signals. Analysis may be done in real time or later, when the signals have been stored on tape.

The signal magnitude may be too large or too small for the filter. An attenuator and/or an amplifier may be required, but are not shown in Fig. 7.5.

The "filter" is the heart of any analyzer. This is an electronic circuit that "passes" ac signals within a relatively narrow frequency band and that "stops" or rejects others. If there were no filter in Fig. 7.5, the ac-to-dc converter would measure a summation of all signals present, as is done with conventional ac-sensing electronic voltmeters.

past a fixed frequency (at which there is a signal component), first a little signal, then more through the peak, then less as the filter is tuned away. Narrow filters more closely approximate line spectra, but the base is always broadened.

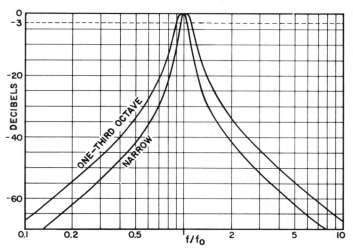

Fig. 7.7 Many analyzers per Fig. 7.6 have two filter widths: broad for finding a frequency component and narrow for more precisely identifying an f_n. Bandwidth often refers to -3 dB or "half power" points.

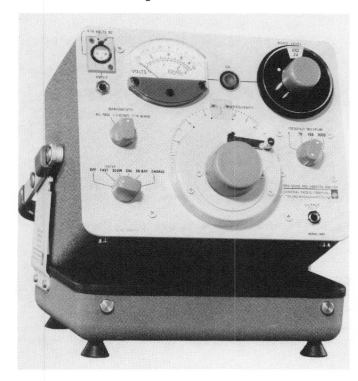

Fig. 7.6 Manually tuned analog spectrum analyzer, suitable for studying vibration, sound, etc. Courtesy GenRad.

On manually-tuned analyzers similar to Fig. 7.6, a dial coupled to the tuning knob indicates f_o. The control can be motor driven for automatic sweeping. Modern analyzers, such as in the system of Fig. 7.8, are electronically tuned; frequency is read from an electronic frequency counter.

Fig. 7.7 shows dual selectivity characteristics. Relatively broad selectivity is used to locate a component; then the narrow bandwidth is used to more precisely identify its frequency. f_o, the filter's center frequency, is tuned much like a radio receiver. A readout meter indicates the peaks.

7.5 Filter bandwidth. Bandwidth of analyzers suggested by Fig. 7.6 and 7.7 varies with f_o. 1/10 or even 1/3 octave is satisfactory at low f_os. (At 10 Hz or 600 RPM, 1/3 octave is about 3 Hz while 1/10 octave is about 1 Hz.) But greater selectivity is needed at higher f_os, say 1,000 Hz, when 1/3 octave is about 300 Hz. Crystal or magnetostrictive filters with fixed bandwidths of 1, 2, 5, 10, 50 etc. Hz are used in the unit of Fig. 7.8. Very narrow filters are needed when several frequencies occur close together. One source may "mask" another, and the true complex spectrum may be obscured. Note the X-Y plotter.

7.6 "Line" spectra look otherwise, due to filter limitations. "Line" spectra, (Fig. 7.1 and 7.2), are theoretically correct. But they often look more like Fig. 7.7 and 7.10. As a filter is tuned

Fig. 7.8 Whereas filter bandwidth on units per Fig. 7.7 vary with f_o, the bandwidth of the analog "tracking filter" (second from bottom unit) remains constant. Oscillator (bottom unit) is programmed to vary f_o. Note X-Y plotter at top and log converter beneath it. Courtesy Scientific-Atlanta.

7.7 Meter readout. An ac signal can be taken from the filter of Fig. 7.5 to, say, a frequency counter. However, Fig. 7.5 shows the ac filtered signal converted (or "detected" or "rectified") to a dc signal measuring its intensity. Readout may be a meter movement or a Y deflection per Sec. 7.8.

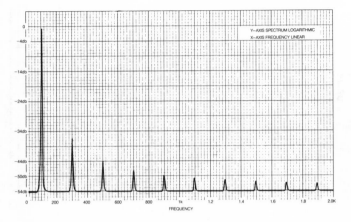

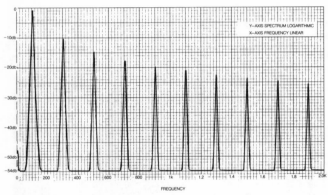

Fig. 7.9 Visualize how a 100 Hz square wave would appear on a 'scope in the time domain. Here are two spectral views in the frequency domain, both with linear frequency scaling. The upper graph shows linear Y scaling, with classical 1, 1/3, 1/5, 1/7, etc. amplitudes at 100, 300, 500, 700, etc. Hz, while the lower shows logarithmic (decibel) voltage scaling, with peaks at 0, -9.54, -13.98, -16.90, etc. dB. Plots were made on analyzer of Fig. 7.8. Courtesy PACCAR.

7.8 X-Y plotting of spectra per Fig. 7.9 gives a permanent or "hard copy" record. Here acceleration vs. frequency was plotted. X pen motion was controlled by the varying f_0. Where significant vibration existed, significant signal passed the filter, was converted to dc and deflected the pen upward.

7.9 Multi-band analyzers, mentioned in Sec. 25.2.2, should also be mentioned here. In principle they resemble the multi-reed mechanical analyzer of Fig. 7.4. Fixed filters simultaneously receive an ac input signal. Certain individual filters will respond—those within whose bands a signal component exists. The readout can be a series of meters per Fig. 25.4. Alternately, a spectral presentation (per Fig. 25.5) is achieved by repetitively stepping quickly through the detector outputs of Fig. 25.4, to indicate the signal magnitudes within the various bands.

7.10 FFT and hybrid (real time) analyzers. Nearly all spectrum analyzers today use digital techniques and the FFT. Digital computers require appropriate programs or software *and* a skilled operator. Hybrid (part digital, part analog) computers lack some flexibility, but are easier to operate. They are often called Real Time Analyzers or RTAs.

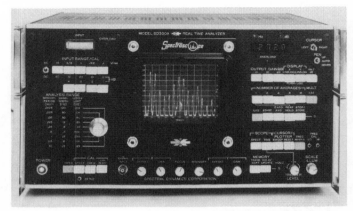

Fig. 7.10 Portable Real Time Analyzer (RTA). Courtesy Scientific-Atlanta.

An RTA is a very fast stepped-band analyzer whose frequency-tuned bandpass filter converts an input signal from the time to the frequency domain, so quickly it is said to operate in "real time."

RTAs contain three basic sections: 1) the input signal conditioner, 2) the time compressor and 3) the Fourier Analyzer, per Fig. 7.11.

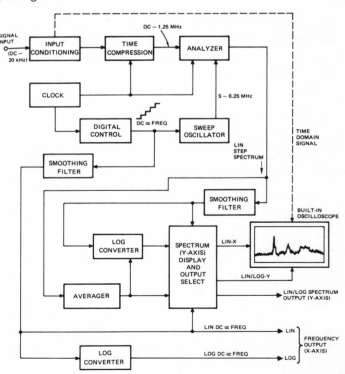

Fig. 7.11 Block diagram of typical RTA. Note three major sections: input signal conditioner, time compressor and Fourier analyzer. Courtesy Scientific-Atlanta.

RTAs use digital time compression to prepare the input signal for Fourier analysis. The time compressed signal is frequency analyzed using typically 250 or 500 synthesized filter "windows," that are tuned by a built-in sweep oscillator. If our upper frequency f_u were, say, 2,000 Hz, each of 500 windows would be 4 Hz wide. We can say that our resolution then is 4 Hz. Some analyzers require accessory oscilloscopes, ensemble averagers, etc. Others, per Fig. 7.10, have these features built in. X-Y recording is external. RTAs are often incorporated in digital vibration control systems (see Sec. 25.2.2).

7.10.1 The input signal conditioner contains anti-aliasing filters and supplies typically 5 volt RMS band-limited signals to the time compression circuits. The filter BW and upper frequency limit f_u are set when you front panel select a frequency range. This also controls the sampling rate in the time compression circuits.

7.10.2 The time compression section includes sample-and-hold circuits, an analog-to-digital converter (ADC), a recirculating memory and finally a digital-to-analog converter (DAC).

7.10.3 The analyzer section is similar to swept-band analyzers, except that an RTA operates at a much higher speed and f_o. It converts from the time to the frequency domain.

How would a 600 Hz sine wave be analyzed by an RTA? First its signal magnitude is adjusted to 5 volts and then it is sampled at three times f_u. Here we'd use f_u = 2 kHz and sample at 6 kHz. Each sample's instantaneous amplitude is ADC'd to a digital 8-bit word. The ADC's output, therefore, is 6000 8-bit digital words per second.

They go into a recirculating memory. When the memory is full, they are outputted at 3.75 MHz, then impressed upon the DAC (which reconstructs the waveform); effective speed = 625 times (3.75 MHz divided by 6 kHz). Various f_us have different speed up ratios. Our 600 Hz sine wave, sped up 625 times, is now at 375 kHz.

The time compressor is essentially a digital frequency multiplier with an analog output. This output waveform may vary from near 0 to 1.25 MHz on all frequency ranges, depending on the frequencies at the input. 600 Hz is found at 375 kHz. 2,000 Hz would be at 1.25 mHz (2,000 Hz x 625). This relationship exists on all frequency ranges as speedup time varies; we will always have 0 to 1.25 MHz from the time compression unit.

In the analyzer section, data from 0 to 1.25 MHz is mixed with a sine wave sweeping 5 MHz to 6.25 MHz, then jumping to 5 MHz to begin another sweep. Any signal in the band of interest passes through a 5 MHz wide crystal filter centered at 5 MHz, then to the output detector, which displays the amplitude and frequency of all input signal components, a complete Fourier analysis of the input waveform. The VCO drive voltage also provides a frequency reference for X-axis plotting.

7.10.4 Summary. Conditioned data first enters the compression circuit, which samples the input data and shifts it to a high frequency. It mixes with a 5MHz to 6.25 MHz sweeping sine signal. After filtering it passes to the output detector and

the display Y axis to indicate the Fourier components of the input signal.

7.10.5 Applications of RTAs include periodic signals, rotating machinery analysis, acoustic analysis, transient signals and random vibration analysis. (For more detailed information on a particular analyzer, consult the manufacturer's service department.)

7.10.6 Random vibration analysis demands an estimate of the power spectrum (PSD). This involves averaging a squared value of a signal (within a narrow bandwidth) over a period of time. The longer the averaging period, and the smaller the bandwidth, the closer we approach the true PSD.

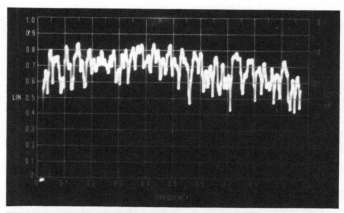

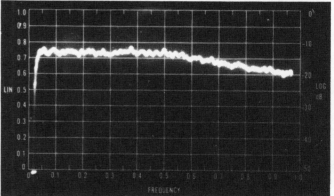

Fig. 7.12 Random vibration test spectrum, without averaging, (upper). Successive pictures of CRT would differ greatly. Lower photo shows effect of averaging; this spectrum is often X-Y plotted.

Such an averager examines successive spectra and computes the sum, averaged over a pre-determined length of time. Fig. 7.12 shows a typical random spectrum before and after averaging. The averaged display is often plotted on an X-Y recorder.

The accuracy of that average is governed by a parameter known as N, statistical degrees of freedom (DoF). The N associated with a section of data relates to the number of samples required to reconstruct a data signal. Nyquist showed that N is directly related to the signal bandwidth, and that 2BT samples are sufficient. B is the data signal bandwidth in Hz. T is the time in seconds over which a data sample is examined. DoF = N = 2BT. Each RTA sweep produces a spectrum estimate having 2DoF. Any desired statistical accuracy can be achieved by averaging over a sufficient number of memory periods.

Note that N as used here is an arbitrary symbol and is *not* the same N used elsewhere to represent force in newtons.

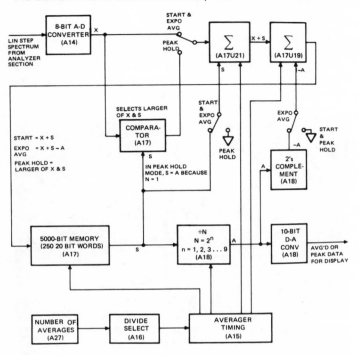

Fig. 7.13 Block diagram of typical signal averager, an important part of any RTA. Courtesy Scientific-Atlanta.

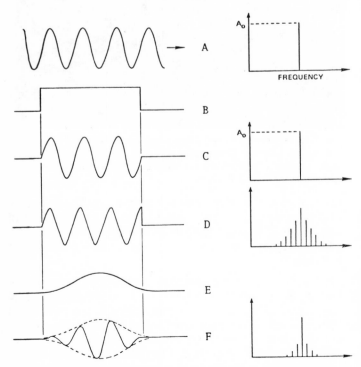

Fig. 7.14 Effect of "windowing" the incoming signal.

Fig. 7.13 block diagrams a typical averager. You will select the averaging mode (see Sec. 7.10.7). If you front panel select 64 ensembles you will have 128 DoF. The digital information in memory is DAC'd and then displayed. Choose between linear and logarithmic on both X and Y scaling.

7.10.7 Averaging modes. RTAs usually offer several averaging modes: Exponential, Linear, and Peak-Hold. Use exponential mode for random analysis; it is equivalent to analog RC averaging. The average is updated once every analyzer memory period. Updating continues until the operator pushes a button; this stores the data in memory. Linear mode loads a preselected number of ensembles into memory and averaging stops; the spectrum is displayed. In the Peak-Hold mode the output displays the maximum values of the averaged signals within each of the say 500 windows. These values are accumulated continuously and compared with accumulated ensembles. Only when and where new data exceeds the accumulated ensemble will it be displayed.

7.10.8 Windowing aids RTAs in displaying a "true" line spectrum. A spectral line represents, in the time domain, a continuous sine wave, as in Fig. 7.14 (A). If the "gate" or "sampling period" at (B) passes an integral number of cycles, as at (C), the spectral display shows the correct single line. But if the gate is slightly "off" per (D), the spectral display shows additional "ghost" lines, subject to misinterpretation. The solution: windowing per (E), a slow opening and closing of the gate. The effect, per (F), is a lessening of the "ghosts."

7.10.9 Random vibration testing. Although RTAs lack the flexibility of "all digital" signal processors and computers, they are exceedingly popular. Independent RTAs provide immediate display of manual equalization (see Sec. 25.2.2). Hybrid analyzers save time, are easy to operate, are portable and inexpensive.

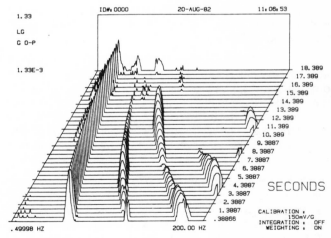

Fig. 7.15 A "waterfall" display, a regular series of spectra, with the latest spectrum being written at the top and appearing to slowly move down the CRT screen. Similar displays are made while rotating machinery speed changes, with spectra plotted at discrete RPM.

7.11 Advanced applications. RTAs can perform a wide variety of tasks, such as (when incorporated in complex systems) monitoring machinery vibrations per Sec. 8.5. The task suggested by Fig. 7.15 once required an additional minicomputer. But now advanced RTAs include various useful, specialized, capabilities.

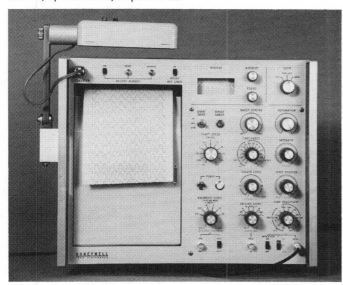

Fig. 7.16 A recording oscillograph. Courtesy Honeywell.

7.12 Oscillographs as in Fig. 7.16 give us a permanent time history of various signals. "Direct writing" o'graphs record low-frequencies (up to perhaps 100 Hz). Electrical signals deflect a pen which makes traces on paper. These records may be used immediately. Up to about 5,000 Hz, the electrical signals deflect mirror galvanometers which reflect small spots of light onto photographic paper. This must be processed before viewing. Recording paper is motor driven at selected speeds. Why not use an oscilloscope and camera? It is ideal for recording one or two traces, for a brief time interval, say shock pulses which contain much high frequency energy. By contrast, a multi-channel recording oscillograph can clarify the time relationship between various events. Recording oscillographs may have up to 30 channels and, at low paper speeds, may record for hours. Each instrument has its area of usefulness.

Neither is much help for getting useful information from complex or random vibrations and complex shocks. Far more useful information is gained from the frequency-analyzing instruments described earlier.

7.13 Tape recording is very convenient for storing and recovering electrical signals. Recovered (played back) signals (hopefully unchanged from what was originally recorded) can be measured, analyzed, viewed on a 'scope, etc. at a convenient time. In environmental test labs, during shock and vibration tests, we often record input motions and forces, also response motions. Often we erase the tape without using it. But imagine that we've had a specimen failure that no one can explain. We'll check the taped inputs and responses just before that failure occurred. Or imagine that our shaker "dumped;" we want to know what inputs and responses resulted. With the information on tape, we can find out. In analyzing rotating machinery, we can carry a portable tape recorder into the plant, then later perform spectrum and other analyses "off line" (via tape playback) in our laboratory.

Section 8
Machinery Health Monitoring -
an Application of Spectrum Analysis

8.0 Introduction. Sec. 8 applies mainly to rotating and less to reciprocating machinery. Its secondary target: those who specify, check performance or operate or maintain machinery. Its primary target: environmental, flight or field test personnel (who work with vibration) who may need to apply material from other sections to machine vibration measurement and analysis.

Excessive vibration prematurely (before wearout) indicates degradation from a machine's original precision. Bearing wear, shaft misalignment, loosened or worn or broken parts increase vibration levels, especially if resonances of machines, bases, pedestals, supports and/or pipes are excited. Eventually a catastrophic failure can occur, with very high costs, also lost revenue. The Vibration Institute and other authorities state that three major difficulties (unbalance, misalignment and bearing instability) cause 80% - 95% of all machinery problems.

"Breakdown maintenance" (running machinery till it destructs), is often very costly. "Preventive maintenance" (periodic overhaul based upon time or usage) can also be expensive, and the overhaul often *creates* damage. Condition monitoring for "predictive" or "on demand" maintenance is least expensive. Take the machine "off line" at a convenient time after you diagnose a difficulty. You will have parts ready, minimizing downtime.

8.0.1 Instrumentation better than human senses. With today's increased operating RPMs and smaller x and $\dot{x}$, human senses are less useful than formerly. Personnel seldom recognize gradual changes. Finally, certain hazardous machinery locations are "off limits" to personnel.

8.0.2 V or A often better criterion than D. Highly reliable displacement probes (see Sec. 3.2.1) are permanently installed on critical machines having high casing/rotor mass, to sense any "wobbling" of a shaft due to unbalance or misalignment. Note that overly simple x limits should apply only to low speed machines, vibrations to say 1,000 cpm. Probe data complements data from velocity sensors (Sec. 4.1), say 1,000 to 60,000 cpm, or accelerometers (Sec. 5.5 and 5.8) above say 60,000 cpm. These sensors are often permanently mounted on bearing supports on critical machines having low casing/rotor mass. Or portable instruments as in Fig. 8.1 may be used. (Permanently affix attachments to

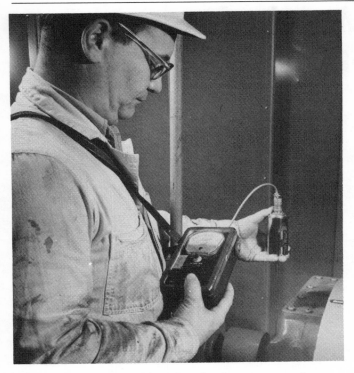

Fig. 8.1 Regular "overall" vibration checkups can reveal gross increases of vibration level, indicating need for further investigation. Courtesy of Marbon Chemical and IRD.

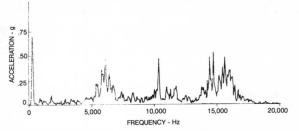

Fig. 8.2 Steam turbine acceleration frequency spectrum. Courtesy Dymac.

8.1 Need to interpret spectra. If checks per Fig. 8.1 show that overall vibration has increased significantly, more sophisticated investigation into precise cause is needed. Table 8.1 lists common vibration mechanisms. f_fs are multiples of the machine's RPM, to identify specific troubles. Visualize the multiplicity of f_fs when several shafts run at different RPM, each dynamically unbalanced, misaligned and/or bent, when bearings and gears are damaged or loose.

Use of Table 8.1 requires spectrum analysis. We need to know how much vibration existed previously (a "baseline" spectrum) and how much exists now, at various discrete (see Sec. 7.2) frequencies. See AN INTRODUCTION TO MACHINERY MONITORING AND ANALYSIS by J. S. Mitchell, PennWell Publishing, Houston.

8.2 Why machines vibrate. Some of the mechanisms mentioned in Table 8.1 are briefly discussed here.

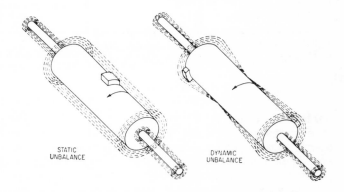

Fig. 8.3 Static (possible to detect without rotation) rotational unbalance forces on supports are in-phase. However, a statically balanced rotor may be dynamically unbalanced, undetectable without rotation. Forces on supports are out-of-phase.

always emplace sensors at identical locations.) Modern practice limits $\dot{x}$ or $\ddot{x}$. Per Sec. 2.4 and 2.6, microscopic x at high f_fs give high $\ddot{x}$ and forces. Velocity sensors are useful across the entire spectrum. With signal integration (see Sec. 3.2.3 and 4.4) they emphasize low f_fs and (rarely) with differentiation (see Sec. 4.5) they emphasize high f_fs.

8.0.3 Trend toward using accelerometers. Imminent failure of rolling element gears and bearings, producing high frequencies, is best sensed by accelerometers. (See Sec. 5.) Cavitation, ball bearing vibrations, gear meshing, etc., also reciprocating engine detonation or "pinging" create high vibration f_fs. Readout may be in terms of $\ddot{x}$ or $\dot{x}$ or (especially at low RPMs) x.

A recent development: PE accelerometers sensing acoustic emission into the megahertz region, giving "shock pulse" information and guarding against bearing and gear failure and against structures cracking. Bandpass filters (typically 80-120 kHz) ignore low f_f disturbances (e.g. unbalance forces) and suffer many fewer "false alarms."

8.0.4 Overall value has limited usefulness. Information presented as "overall" magnitude is adequate for installed-sensor monitoring of less critical machinery. If it increases, consider Sec. 8.3. Note that a major increase in a small spectral peak of Fig. 8.2 would not measurably increase the "overall" reading. On critical machinery, spectral monitoring provides more information, if decision-making personnel can interpret it. The Canadian Naval Engineering Test Establishment and others more easily use octave band analysis; most workers want 1/3 octave or less width.

8.2.1 Rotational unbalance and balancing. If rotor mass is centered along its axis, rotation is smooth. But any off-axis "heavy spot" creates first-order or fundamental vibration at f = RPM/60 Hz, noted at bearing supports. Fig. 8.3 suggests the forces, at shaft speed or "first order" or "1 × RPM" frequency per Table 8.1. Magnitude depends upon amount of unbalance, upon RPM, also upon shaft and bearing support resonances. Cavitation may have eroded a pump impeller. Material may have adhered to an impeller, then broken off unevenly. A grinding wheel may have worn unevenly. A turbine may have "thrown" a blade. A coupling may be worn.

Vibration Identification Chart

Cause	Amplitude	Frequency	Phase	Remarks
Unbalance	Proportional to unbalance. Largest in radial direction.	1 × rpm	Single reference mark.	Most common cause of vibration.
Misalignment couplings or bearings and bent shaft	Large in axial direction 50% or more of radial vibration	1 × rpm usual 2 & 3 × rpm sometimes	Single double or triple	Best found by appearance of large axial vibration. Use dial indicators or other method for positive diagnosis. If sleeve bearing machine and no coupling misalignment balance the rotor.
Bad bearings antifriction type	Unsteady—use velocity measurement if possible	Very high several times rpm	Erratic	Bearing responsible most likely the one nearest point of largest high-frequency vibration.
Eccentric journals	Usually not large	1 × rpm	Single mark	If on gears, largest vibration in line with gear centers. If on motor or generator, vibration disappears when power is turned off. If on pump or blower, attempt to balance.
Bad gears or gear noise	Low—use velocity measure if possible	Very high gear teeth times rpm	Erratic	
Mechanical looseness		2 × rpm	Two reference marks. Slightly erratic.	Usually accompanied by unbalance and/or misalignment.
Bad drive belts	Erratic or pulsing	1, 2, 3 & 4 × rpm of belts	One or two depending on frequency. Usually unsteady.	Strob light best tool to freeze faulty belt.
Electrical	Disappears when power is turned off.	1 × rpm or 1 or 2 × synchronous frequency.	Single or rotating double mark.	If vibration amplitude drops off instantly when power is turned off, cause is electrical.
Aerodynamic hydraulic forces		1 × rpm or number of blades on fan or impeller × rpm		Rare as a cause of trouble except in cases of resonance.
Reciprocating forces		1, 2 & higher orders × rpm		Inherent in reciprocating machines, can only be reduced by design changes or isolation.

Table 8.1 From MECHANICAL ENGINEERING, October 1969. We recommend Wavetek Rockland's highly detailed "Machinery Vibration Diagnostic Guide" wall chart. It is more complete and more recent than Table 8.1.

Factory balancing machines, usually much larger than Fig. 8.4, are used for production. Rotors are emplaced, spun, and unbalance forces suggested by Fig. 8.3 are noted. Mass is manually or automatically added or removed at appropriate locations.

Fig. 8.4 Production balancing machine, courtesy B&K.

Field or in-place maintenance balancing, also called trim balancing, of installed machines is done without removing rotors, using specialized portable equipment suggested by Fig. 8.5 (can be much more elaborate). However, you could use your digital or hybrid spectrum analyzer (Sec. 7.10 to identify vibration sources.

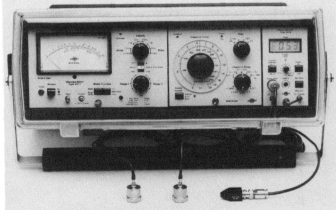

Fig. 8.5 Typical battery operated portable unit for field balancing. Courtesy B&K.

8.2.2 Shaft misalignment refers to improper bearing geometry, usually between separate, coupled machines. It may result from faulty installation, or thermal growth, or one machine can shift.

8.2.3 Bearings, if defective, can cause serious vibration problems. Rolling contact bearing failures give higher order vibration. Sleeve bearings can develop instabilities at a partial order. See Fig. 0.3.

8.2.4 Gears produce vibration at relatively high frequencies, per Table 8.1. Tooth failure or other rough meshing intensifies high f_f vibration.

8.2.5 Resonance greatly magnifies any of these vibrations if an f_f coincides with an f_n of the machine or its supports, per Sec. 1.1.2.

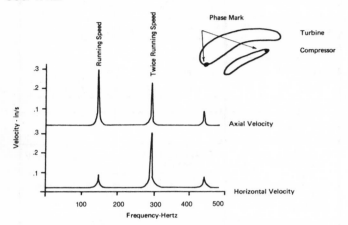

Fig. 8.6 A typical misalignment signature plot. After Borhaug.

8.3 Identifying sources of vibration. Spectrum analyzers tell us the frequency of spectral changes. We easily relate these to specific mechanisms on simple machines, having few rotating components. But on a complex machine there are many mechanisms at work, overwhelming us with too much data. Try listening to or touching or watching various parts of that machine. Use a "strobe" light tuned to $f_f \pm 1$ Hz; a faulty part seems to move at 1 Hz so you can *see* what is happening. See Sec. 16.9. A newer approach: video recording, yielding instant replay.

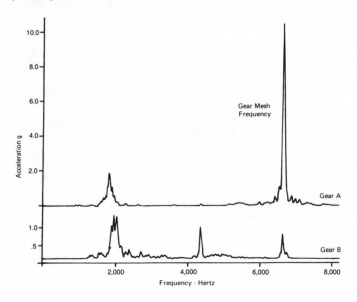

Fig. 8.7 Speed increasing gear high frequency signature. After Borhaug.

8.3.1 Analog analysis. Use a tuned analog analyzer per Sec. 7.4. Tune it manually, and hand log resulting data. Better, sweep frequency with a programmed oscillator. Best, an x-y graph gives a permanent record. With a narrow filter, great spectral detail is possible, given sufficient time.

8.3.2 Shaft speed tunes analyzer. Use a tachometer or light reflections onto a photocell (compensating for brief RPM changes during analysis) to automatically tune your filter for study of a specific vibration component (such as a particular shaft unbalance) at a specific order throughout the machine's speed range.

8.3.3 Digital analysis and control. Signals may be digitized and then FFT'd into spectral information, per Sec. 7.10. Great spectral detail, which with analog analysis and plotting per Fig. 7.8 took much time, is almost instantly displayed on a CRT, per Fig. 7.10. On the screen of a multi-channel computer, you can visually compare current vibration spectra with "base line" spectra.

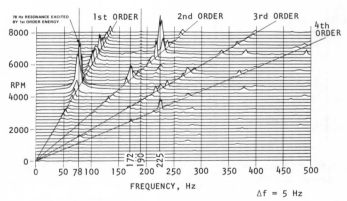

Fig. 8.8 RPM spectral map or dynamic Campbell diagram. Courtesy H-P.

Fig. 8.8 shows a series of plots comparable to Fig. 8.2 but taken very rapidly as a machine swept through its RPM range. Resonances or critical f_fs appear as peaks along the "order" lines.

8.4 Storage on magnetic tape. Narrow band analog spectral plotting (as in Fig. 8.2) often requires 10-30 minutes, during which operating conditions (such as RPM) may change. Lacking rapid digital equipment just described, you *could* save much field time by tape recording. Voice annotate each tape segment concerning pickup locations and what happened. Later, duplicate important segments onto a tape loop, then analyze spectra.

8.5 Warning systems. Machines can self-destruct before circuit breakers and bearing temperature monitors can respond. Fortunately, vibration warning systems respond quickly, long before catastrophic failure.

Most published data unfortunately is "overall" or "wideband." It is difficult to specify limits (constant x, constant ẋ or

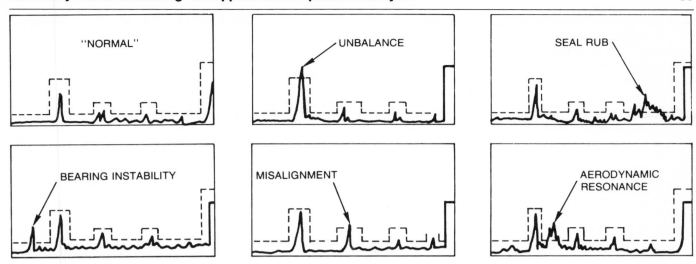

Fig. 8.9 Examples of spectral exceedances above "windowed" spectrum. Courtesy Dymac.

constant $\ddot{x}$) that apply at all frequencies. See VIBRATION TOLERANCES FOR INDUSTRY, by Baxter and Bernhard of IRD, ASME publication 67-PEM-14. See also MACHINERY VIBRATION STANDARDS by Eshleman of The Vibration Institute.

Better than "overall" control: trigger when the existing spectrum (narrowband analysis) exceeds a "windowed" spectrum, as in Fig. 8.9. Quite likely none of those changes would much increase the overall vibration level. Guarding only important f_f windows decreases chances of a "false alarm" and permits high sensitivity to real difficulties.

8.5.1 Engineering analysis. Establishing limits requires several steps. First, per Table 8.1, calculate f_fs to guard. Then, using guidelines or your own experience, select several monitoring points and sensors.

8.5.2 Obtain baseline data. To properly evaluate future spectral changes, survey vibration while the machine is operating properly, ideally when it first goes into service. Start up, run at various speeds and shut down, meanwhile recording data and, in great detail, exactly how that data was taken. Plot spectra on portable equipment taken to the job site, or store vibration signals on magnetic tape (per Sec. 8.4) for later plotting.

8.5.3 Set limits. Perhaps the critical machine's manufacturer has set limits. Or your protection equipment manufacturer provides limits. Perhaps you can find other guidelines. Best: your own experience. Have similar machines previously failed? Use their data. Or guess at limits; later observe operating vibration and change your limits as you accumulate experience.

8.5.4 How much monitoring? Monitoring schemes range from a single channel meter and alarm to highly sophisticated computer controlled systems having hundreds of channels. Bentley-Nevada, Dymac, IRD, Metrix, PMC-Beta, Palomar and others will send you catalogs.

Figs. 8.10 and 8.11 suggest benefits of spectrum analyzer + trend monitor + computer. One system automatically moni-

tors up to 256 channels such as bearing vibration, rotor axial position, gear mesh conditions and seal rubbing, also oil

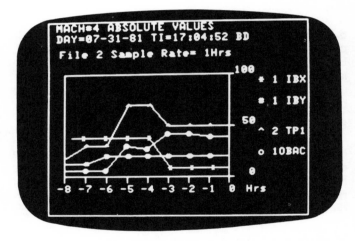

Fig. 8.10 Trends for last 8 hours, 4 sensors. Courtesy Dymac.

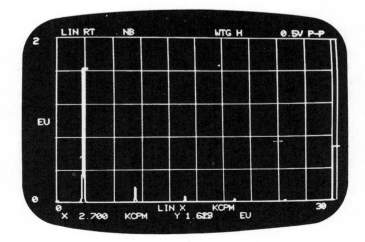

Fig. 8.11 Vibration spectrum at one sensor. Courtesy Dymac.

and bearing temperatures. It reports current information for human interpretation and stores historical data for establishing trends. Data is instantly available as bar graphs, trend graphs and detailed comparison of spectra. Your operators can base decisions on current vs. historical vs. baseline conditons.

8.6 Decision aids. A warning system can alert a machine's operator. Then what? He needs help in deciding what to do. His "expert" might not be available. Alone, he must identify the difficulty, decide if it is serious, and what action to take. One system combines a digital spectrum analyzer and a computer programmed with a "confirmation matrix." It verifies and correlates various machine measurements (including vibration) in a few minutes, then tells the operator what to do and how soon.

Diagnosis is most important during and after startup and during coast-down. Often machines have run well for months but give trouble after a shutdown, during thermal stress. Streamline your data acquisition. Speed is essential.

Most protection is based upon uniaxial vibration sensing. But some machines, especially reciprocating engines and synchronous electric motors, need torsional vibration monitoring. Consider Sec. 29.

8.7 Aircraft engine vibration monitoring (EVM) early applied remote sensing of vital rotating equipment "health". Early jet engines were not yet trusted. EVM was an alternative to frequent overhauls. Should a bird or stone be ingested or a turbine blade fail, the crew shuts down the damaged engine before it shakes itself off the airplane (as has happened).

Early EVM systems, unfortunately, used $\dot{x}$ sensors (see Sec. 4.1), unreliable in jet engine environments. Ground crews often could not confirm the difficulty. Some airlines have disabled their EVM systems. (That some alarmed engines have later failed is not widely publicized.) A benefit of EVM: it permits predictive or "on condition" maintenance, less expensive than regular "time between overhaul" maintenance.

Fig. 8.12 Turbofan engines are vibration monitored. Courtesy Boeing.

Current EVM systems use highly reliable PE accelerometers (Sec. 5.8), some capable of withstanding +1,400°F, mounted on the engine. Some accelerometers contain amplifiers per Sec. 5.13 and/or differential outputs, and use simple twisted pair wiring. Others connect through heavily shielded cables to filters, charge converters, integrators (for readout of $\dot{x}$), ac-to-dc conversion and display as in Fig. 8.13. Typical ranges are 0 to 2 in/sec or 50 mm/s, with alarms that sound at typically 1.1 in/sec or 60 mm/s. An additional integration (for x readout) can be provided. Some systems include a tracking

filter (see Sec. 14.7) automatically tuned by fan or turbine speed, so that only first-order vibration is measured.

If a flight crew notes gradually increasing vibration on an engine, ground personnel can connect a spectrum analyzer to the already-mounted sensors. Diagnosis can lead to minor adjustment (such as fan or propellor rebalancing) or to engine removal and overhaul.

Fig. 8.13 Aircraft vibration instruments. Courtesy Endevco.

Recent EVM systems have "gone digital." Sharp digital filters avoid the instabilities of analog circuits. Digital diagnostics and readout inspire more confidence. Finally, digital processing and automatic recording lessens flight crew labor; maintenance personnel receive history and diagnosis without ground run-ups, saving fuel and labor.

Similar but more elaborate sensors, readout and analysis systems are used for production checks of new or overhauled jet engines. Instruments similar to Fig. 8.14 are found outside many turbine test cells.

These ideas are used in other applications of gas turbine (and, less often, internal combustion) engines and gear trains on helicopters, pumps, compressors, generating stations, etc. Engine condition is ascertained, the defect is diagnosed and service is planned. Long-term trend analysis permits scheduling of maintenance. Analysis leads to quantified redesign goals.

8.8 U.S. Navy shipboard vibration monitoring. Many Navy equipments (particularly aboard submarines) must meet stringent specifications for low vibration levels at their foundations. However, machines are supported by soft vibration isolators during acceptance tests ashore. Vibration levels could change aboard ship; repeat the tests.

Make regular vibration (and noise) surveys. If overall wideband (unfiltered) levels have increased, spectrum analysis indicates the troublesome f_f. Then use a chart similar to Table 8.1 to identify and remedy the physical cause. Outside assistance may be needed.

condition differs from the "working" environment. If you mount a pump on a resilient spring suspension for test, in-service motion will differ, though internal forces are the same. One factor: shaft alignment in service. Thus "permissible vibration" tables are only guides.

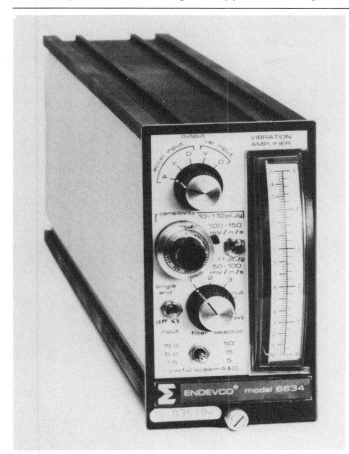

Fig. 8.14 Vibration instrumentation for jet engine test cell. Courtesy Endevco.

8.9 Vibration as a quality control tool. Manufacturers often test their products for vibration. But their "standard" test

Fig. 8.15 Portable vibration measuring equipment in use on production line. Courtesy Consolidated Kinetics Corp.

We express our appreciation to several people for reviewing Sec. 8 and making helpful suggestions: Dr. Jan Borhaug, Ralph Buscarello, John S. Mitchell, Yeve Smith and Mike Walter.

Section 9
Calibration of Vibration Measurement Systems

9.0 Introduction. Before we make any precise physical or electrical measurements, we always calibrate (determine the sensitivity of) our equipment. Sensitivity of a mechanical-to-electrical transducer system (a pickup or sensor + a signal conditioning amplifier + some readout device, perhaps a meter) is usually expressed as the ratio

$$\frac{\text{electrical output signal}}{\text{mechanical input}} \qquad (9.1)$$

Calibration may be described as an orderly procedure for finding this sensitivity. Calibration often involves checking sensitivity over some frequency range (frequency response) and some amplitude range (linearity). Calibration may also include determining phase shift, effects of temperature, etc. We usually adjust amplification so our display is in convenient engineering units. (We will not discuss another definition sometimes found: adjusting pickup sensitivity.)

Accelerometer system sensitivity is usually expressed as

$$\frac{\text{electrical signal in millivolts or picocoulombs}}{\text{acceleration input in g units}} \qquad (9.2)$$

$\dot{x}$ pickup sensitivity would be expressed as mv per in/sec or per mm/s. x pickup sensitivity as volts or mv per inch or per mm. Calibration of force sensors is discussed in Sec. 6.5.

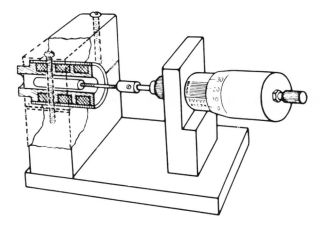

Fig. 9.1 Static calibration of LVDT or any displacement sensor requires precise x increments, obtained by adjusting lead screw at right.

9.1 Static displacement calibration. x sensitivity can be checked statically, without motion. A calibrating device similar to Fig. 9.1 provides relative motion between the fixed and moving parts. Step the moving portion at definite intervals; take voltage readings at each. Suppose that static sensitivity is exactly 1 volt dc per mm. First find the "zero" position (that gives zero output signal). Then note static x increments giving signal increments per Table 9.1 and Fig. 9.2. Sensitivity is the slope of the best straight line through the data points.

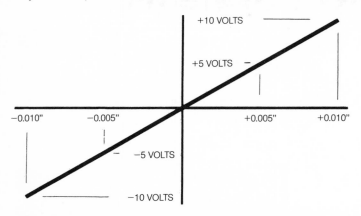

Fig. 9.2 Static sensitivity determination for displacement sensor.

Table 9.1

Static Displacement -mm-	Output -volts-
−.010	−10.0
−.005	− 5.0
0	0
+.005	+ 5.0
+.010	+10.0

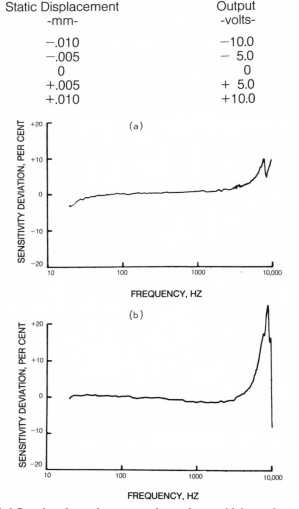

9.2 Static accelerometer calibration. Strain gage and PR accelerometers (as suggested by Fig. 9.3), also servo units, are simple to calibrate. Hold the pickup as shown on a flat, level surface. Balance the electrical output to zero. Now stand the pickup on end for a 1g change. Its new output is the sensitivity per g, say 1 volt. We could use a centrifuge for higher static accelerations.

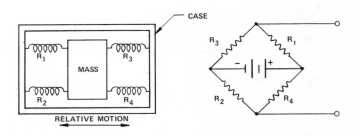

Fig. 9.3 Accelerometer which can be both statically and dynamically calibrated.

9.3 Dynamic determination needed. However, don't use *static* calibration data for vibration measurements without first checking *dynamic* sensitivity over your planned frequency range. (Any pickup or mounting resonance will change its sensitivity.) Attach the pickup to a shaker that moves sinusoidally. Drive the pickup with constant motion amplitude over your frequency range. Plot sensitivity as in Fig. 9.4. If sensitivity is not constant, find out what corrections your data will need.

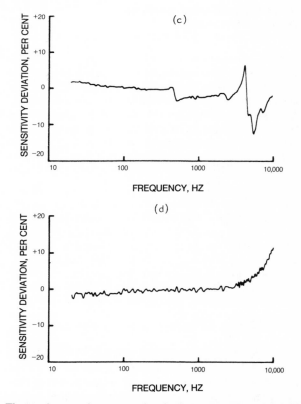

Fig. 9.4 Graphs of accelerometer dynamic sensitivity vs. frequency. These show various non-classical responses, emphasizing danger in only determining sensitivity at one frequency.

There are several methods for determining an accelerometer's f_n. Using one technique or another, check f_n often. It may shift due to dropping or other severe shock, indicating need for repair that may not "show up" during routine calibration at moderate intensity, per Sec. 9.

Fig. 9.5 Single frequency (100 Hz) calibrator delivering one intensity (1g RMS) to an accelerometer. Courtesy GenRad.

9.4 Single-frequency checks. Fig. 9.5 shows a self-contained accelerometer calibrator for in-the-field quick checks on accelerometer channel sensitivity just before and just after an important measurement. While speedy, this will miss abnormalities at other frequencies, as in Fig. 9.4.

9.5 Frequency response is determined with a shaker (per Section 9.9) having very low distortion and good straight line motion. Check pickup sensitivity at say ten frequencies. Better, since you might still miss abnormalities per Fig. 9.4, make a frequency sweep. Use an x-y recorder to plot sensitivity vs. frequency at say three intensities. If all three graphs agree, you have good amplitude linearity. The graphs won't be perfectly flat but you will be able to say, "My pickup is flat within ± ____% over the range ____ to ____ Hz."

To calibrate a PE unit (together with its cable and associated amplifier), mount it properly on an EM shaker/calibrator. Under vibration, our "unknown" accelerometer generates an electrical signal; measure this on a precision ac-sensing electonic voltmeter. Observe its waveform on a 'scope so you can trust the meter (see Sec. 2.7). This measurement becomes the numerator of Eq. 9.1 and 9.2.

9.6 Absolute calibration. Ideally, we will learn sensitivity in terms based upon distance, time and voltage. In practice, comparison calibration per Sec. 9.7 is far more popular.

9.6.1 Optical calibration (generally used from 15 to 100 Hz) involves measuring the mechanical input x in mm or inches D per Sec. 3.1. Measure frequency very closely, since a 1% error here makes a 2% error in our results. Then calculate $A = 0.00202f^2D$ or $A = 0.0511f^2D$, to obtain the denominator of Eq. 9.2. The electrical signal is the numerator. Now calculate

sensitivity per Eq. 9.2. (With velocity pickups, the numerator is usually mv. Quantize the denominator, mechanical input V, as mm/s or in/sec.)

If your measurement channel includes integrators for x readout, check those readouts against optical measurements.

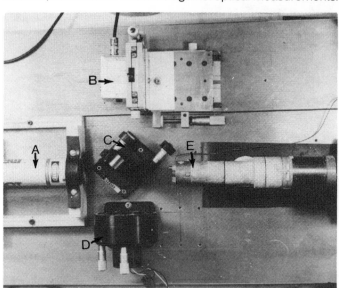

Fig. 9.6 Experimental NBS setup. A is the monochromatic light source. B, the photoreceptor. C, the beam splitter. D, the reference mirror. E, a PE shaker (see Sec. 9.9.2), on which is mounted the accelerometer being calibrated alongside glass cylinders having front reflective coatings.

9.6.2 Interferometry permits extending these ideas to much higher frequencies, where the x to be sensed is much smaller. Difficulties are well summarized in "Calibration of Vibration Pickups at High Frequencies," by B. F. Payne of NBS at the June '83 Transducer Workshop. He describes calibrations to 29 kHz, using X = 121 nm measured by a laser interferometer, with errors estimated at ±2% to 10kHz and ±4% above 15 kHz, using at higher frequencies a PE shaker similar to Fig. 9.18. Fig. 9.6 shows his optical system. Isolation (Sec. 1.1.5) against earth and building vibrations is critical.

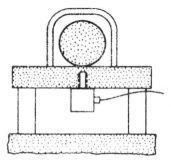

Fig. 9.7 Possible construction of "chatter ball" calibrator.

9.6.3 Chatter calibration was developed by Sir William Bragg in 1919. Accuracy is not high, but its relation to the earth's gravitational $\ddot{x}$ is highly logical. Fig. 9.7 shows the essential parts to be mounted on practically any shaker for calibration at ±1g. So long as A is less than 1g the bronze

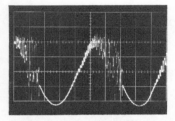

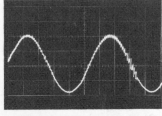

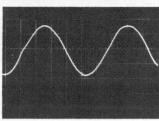

Fig. 9.8 Waveforms from "chatter" calibration (top) about 1.1g peak, (middle) just over 1g peak, and (bottom) just under 1g peak.

Fig. 9.10 Beryllium armature for unit of Fig. 9.9. Standard accelerometer is mounted under the center attachment point. Courtesy Bouche Laboratories.

ball rests on the "floor." The accelerometer's signal (viewed on an oscilloscope) is sinusoidal as in the bottom trace of Fig. 9.8, taken at just under ±1g. If the maximum downward ẍ exceeds 1g, the ball "free falls" momentarily before striking the floor. Impact gives "hash," per the middle trace, taken at just over ±1g. Any readout meters *should* display 1g peak or 0.707g RMS. The upper trace represents about ±1.1g.

9.6.4 Reciprocity calibration involves precise measurements of mass, frequency and resistance, on (usually) two shakers similar to Fig. 9.9. (Fig. 9.10 shows its armature.) Various precise masses are attached, along with three accelerometers (including the one being calibrated). "Reciprocal" calculations are used to calculate the sensitivity of the accelerometer being checked. For good accuracy, the shakers must have low distortion and low transverse motion. The estimated error at 100 Hz is under 0.5%. Further data is found in Sec. 5.2 of the 1979 text SVM-11, "Calibration of Shock and Vibration Measuring Transducers," available from SVIC, Washington, D.C.

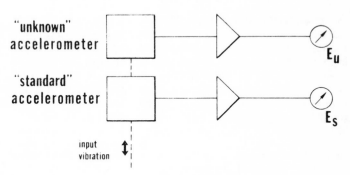

Fig. 9.11 Concept of comparison calibration. See eq. 9.3.

9.7 Comparison calibration is the most popular method of shaker calibrating accelerometers. Two sensors are rigidly coupled together and to a shaker; mechanical and electrical connections are shown in Fig. 9.11. (Sensors must receive identical motions. If in doubt, physically interchange them. If results change, motions were not identical.) The two signals should be proportional to their sensitivities, per Eq. 9.3. Electrical signals E_u and E_s are read, typically, at three different intensities and at ten different f_ss. S_s, the sensitivity of the standard unit, has already been determined by a standardizing agency. The goal is to determine S_u, the sensitivity of the unknown unit, by implementing Eq. 9.4.

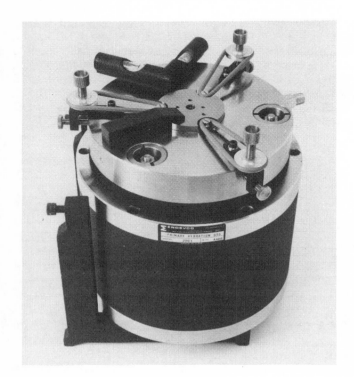

Fig. 9.9 Modern EM shaker/calibrator.

$$\frac{E_u}{E_s} = \frac{S_u}{S_s} \qquad (9.3)$$

$$S_u = S_s \frac{E_u}{E_s} \qquad (9.4)$$

The intensity range should cover the intensities at which the pickup will serve. Especially on engine monitoring duty, $\ddot{x}$ peaks can be many times RMS levels, over long periods.

9.7.1 Back-to-back calibration. We could implement Fig. 9.11 by mounting two accelerometers side-by-side. However, per Sec. 12.2, at certain f_s they would experience large differences in motion. Back-to-back is good to at least 5 kHz. Calibration shaker manufacturers (see Fig. 9.10) attach the standard unit to the armature. Or to a special insert (often ceramic) at table center.

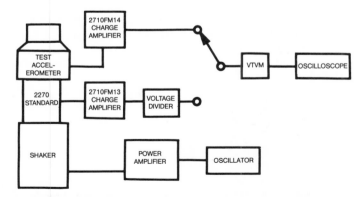

Fig. 9.12 Laboratory standard accelerometer combined with back-to-back fixture to accept "unknown" accelerometer. Courtesy Kistler.

Several manufacturers offer units exemplified by Fig. 9.12, combining a back-to-back holding fixture with a highly stable accelerometer. Together with its own cable and amplifier, it is sent to a standardizing agency for calibration. Upon return, it is used for comparison calibration of other accelerometers.

Fig. 9.13 Block diagram of comparison calibration — "divider method." Courtesy Endevco.

9.7.2 Voltage divider method. Fig. 9.13 requires only one readout meter, and it need not be precise. Assuming that the standard accelerometer has the greater sensitivity, only a precisely adjustable portion of its signal (call this R, for ratio) reaches a selector switch, along with the entire signal from

the unknown accelerometer. The operator adjusts R while switching, until he obtains the same meter reading in either switch position, and then can compute

$$S_u = RS_s \qquad (9.5)$$

Fig. 9.14 Equipment for computer controlled dynamic calibration of accelerometer. Courtesy H-P.

9.7.3 Digital method. The foregoing methods are too slow for laboratories using hundreds of accelerometers. They use random excitation and digitally compute the transfer function between the unknown and the standard units, S_s being already known. Fig. 9.14 shows the necessary equipment. The FFT of E_s (X in Fig. 9.15) is our *input power spectrum* and the FFT of E_u (Y in Fig. 9.15) is our *output power spectrum*.

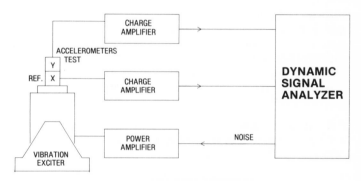

Fig. 9.15 Block diagram — computer controlled dynamic calibration of accelerometer. Courtesy H-P.

The accelerometers are mounted back-to-back. The shaker is driven by an analog signal (from a DAC within the Fourier analyzer). The signal is pseudo random for equal voltage excitation at all frequencies. The input power spectrum is not flat, due to the transfer function of the shaker, the standard accelerometer system characteristics and any resonances in the accelerometer mountings. (There must be adequate energy at all frequencies of interest.) The output power spectrum generally resembles the input power spectrum.

The analyzer first computes the two spectra, then the ratio between them, the transfer function, printed or plotted per Fig. 9.16. Any deviation from unity means that S_u differs from

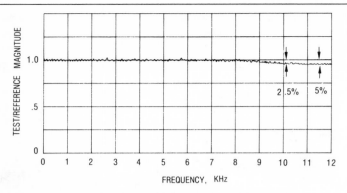

TRANSFER FUNCTION MAGNITUDE

Fig. 9.16 Results of computer controlled dynamic calibration of accelerometer. Courtesy H-P.

S_s. Here we see a unity transfer value up to 7 kHz. If S_s was 10 pc/g and our transfer function value was 1.383, S_u would be 13.83 pc/g. Note the deviation at say 12 kHz. S_u is 5% to 6% less than S_s.

Plotting per Fig. 9.4 and 9.16 prevents use of damaged accelerometers which are abnormal only at certain frequencies. "Ten frequency comparisons" may miss these. Digital calibration can additionally plot phase to check transfer function quality.

With any technique, noise and/or non-linearities introduce errors. The computer can quickly measure these influences by calculating the coherence function to determine how much of the output spectrum is caused by the shaker's mechanical input. For further study, get K. A. Ramsey's 1975 ISA paper "Accelerometer Calibration using Random Noise and Transfer Function Measurements," from H-P.

9.8 Estimated Errors. Many (hopefully small) errors creep into calibration work, even in the best facilities. Examples are listed in Table 9.2.

Table 9.2

ERRORS IN ACCELEROMETERS CALIBRATED BY THE COMPARISON METHOD.

MEASUREMENT	SENSITIVITY ERROR per cent
Reciprocity Calibration Error for Standard, 100 Hz	0.5
Stability of Standard	0.5
Comparison Frequency Response Calibration Error for Standard	
5 Hz - 900 Hz	1.1
900 Hz - 10,000 Hz	2.1
Relative Motion, 900 - 10,000 Hz†	1.0
Distortion	0.2
Voltage Ratio	0.2
Amplitude Linearity – 0.2 g to 100 g	0.2
Range Tracking, Standard Amplifier – 1, 10, and 100 g/ V Ranges	0.2
Range Tracking, Test Amplifier	0.2
Amplifier Relative Frequency Response	0.1
Amplifier Gain Stability, Source Capacity, etc.	0.2
Environmental Effects on Accelerometers, Transverse Sensitivity, Strain, Temperature, etc.	0.5*
Environmental Effects on Amplifiers, Residual Noise, etc.	0.2**
Estimated Error – 100 Hz	1.0***
Estimated Error – 5 to 900 Hz	1.5***
Estimated Error – 900 to 10,000 Hz†	2.5***

  * The error varies from 0% to 0.5% for most accelerometers operated under controlled laboratory conditions.

 ** Applies for controlled laboratory conditions.

*** Determined from the square root of the sum of the squares of the applicable individual errors.

  † Highest frequency is 5000 Hz for test accelerometers with a total mass exceeding 35 grams.

These are calibration laboratory errors. Errors are greater in the field or in the environmental test laboratory. Your errors may approach ±10%. Thus the ironic statement: "We do not measure vibration; we only estimate it."

9.9 Calibration shaker. While you can calibrate on nearly any shaker, Sec. 9.3-9.8 should convince you of need for specialized equipment.

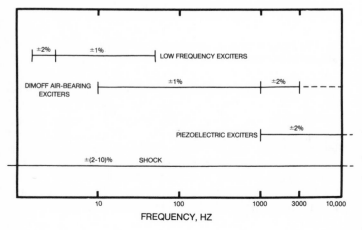

Fig. 9.17 Summary of accelerometer calibration methods and attainable accuracies. Courtesy NBS.

Fig. 9.18 Piezoelectric motion generator for high frequency calibration. Courtesy Wilcoxon.

9.9.1 Electromagnetic calibrators. The EM calibrator of Fig. 9.9 resembles the EM shakers of Sec. 12. Several details differ. An air bearing restrains motion to a straight line. The beryllium armature is light and stiff, its first f_n about 50 kHz. Before assembly, a "known" PE unit, (traceable to NBS) is installed in the armature. Later, when the calibrator is used, various "unknown" accelerometers will be attached.

9.9.2 Piezoelectric calibrators. The NBS piezoelectric (PE) shaker of Fig. 9.6 is similar to Fig. 9.18. At its heart is a stack of crystals that change dimension (very short stroke) when suitably driven.

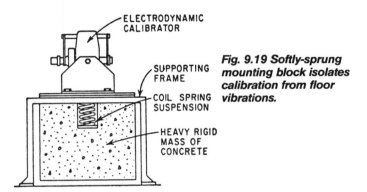

ELECTRODYNAMIC CALIBRATOR

SUPPORTING FRAME

COIL SPRING SUSPENSION

HEAVY RIGID MASS OF CONCRETE

Fig. 9.19 Softly-sprung mounting block isolates calibration from floor vibrations.

9.9.3 Calibrator mounting. The sensor being calibrated should not receive any motion except the precisely-measured calibrator input. Any input from the building must be blocked, usually by a spring-mass suspension as suggested by Fig. 9.19, having a low (1 or 2 Hz) natural frequency. A further caution: extremely sensitive pickups may indicate as vibration any airborne noise. Soundproofing the calibration room is often necessary.

Fig. 9.20 Sensitivity "quick check." After Baber.

9.10 Less formal procedures. In test labs and in the field, check every measurement channel at least weekly, better daily, sometimes before and again after an important test. Mount several sensors on a special fixture, per Fig. 9.20. Accurately inscribe an "optical wedge" or use a convenient "stick on" wedge per Sec. 3.1.3. (Lateral sensors help to

avoid any f_f where your shaker has much lateral motion.) Use one sensor only for calibration. Vibrate the cluster at a D of say 0.1 inch or 2.5 mm, resulting in a useful A per Eq. 2.11 or 2.14. All readouts should agree; if any differ, find out why.

Check flatness of frequency response by sweeping f_f while holding A constant. (Let the "standard" sensor act as "control" for your servo.) Observe signals from all sensors on a multi-channel oscillograph (Fig. 7.16), with paper speed about 0.1 inch or 2.5 mm/second. Set gains on all channels the same: to give record widths about 1 inch or 25mm peak to peak; these should stay reasonably constant while f_f sweeps say 10 - 2,000 Hz. If any widths change, find out why; be sure, for example, that no channel is filtered. Or that none of your sensors is damaged. Keep the oscillograph record to answer the challenge "Did you dynamically calibrate?"

9.11 Extreme temperature calibration. Most laboratory calibrations are done at room temperature. But accelerometer sensitivities change per Sec. 5.16.1. Before using sensors at temperature extremes, calibrate them at those extremes. Most labs place an insulated box over the calibrator/shaker. A fan circulates conditioned air inside. The shaker table is extended inside the box; its axial and lateral motions unfortunately limit such calibrations to 100 Hz.

9.12 Linearity checks. Our discussions thus far have involved relatively low vibration intensities, generally under 100g. For use at higher intensities, check linearity (determine that its sensitivity is constant). For sustained single-frequency sinusoidal vibration at 100g and above, attach it to your shaker via a series of resonant beams per Fig. 19.19.

9.13 Shock calibration is discussed in Sec. 32.4.

9.14 Purchasing calibration services is an alternative to performing your own work. Perhaps you can send your pickups and conditioning amplifiers (as sets) to NBS at Washington, D.C. 20234, or to your own country's standards laboratory. NBS offers (but generally only to certain qualified users) a range of highly accurate vibration calibrations from 2 to 13,000 Hz at fees approaching $2,000. A typical turn-around time is 45 days. You may wish to obtain NBS Special Publication 250. Most accelerometer manufacturers offer recalibration services at more reasonable speed. And so do many of the commercial test labs. Within most military services is a "central" calibration lab, which sends its transfer standards to NBS. Each service's "working" laboratories sends equipment here for calibration, rather than to NBS.

9.15 Calibrate entire system. Note that leading calibration facilities prefer to check entire systems, rather than individual sensor + amplifier + read out, etc., though the latter is sometimes necessary.

We very much appreciate the assistance of B. F. Payne of NBS and P. S. Lederer of Wilcoxon.

Section 10
Mechanical Shakers

10.0 Introduction. During World War II, when vibration damage to airborne equipment first became a severe military problem, engineers used available mechanical shakers to determine equipment ability to operate in spite of vibration. By 1980, mechanical shakers had become historical, as they can't perform random vibration tests. However, requirements such as Federal Test Method 101B and ASTM Standard 999-68 will doubtless continue for some years.

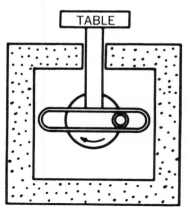

Fig. 10.1 "Scotch yoke" mechanism for driving "brute force" mechanical shakers. It attempts to maintain constant D.

10.1 Direct-drive shakers per Fig. 10.1 and 10.2 provide a table (to which one fastens the test item), driven by a rotating wheel and eccentric. Unfortunately, motion contains much second-harmonic distortion, as well as some "hash" from belts and bearings.

Fig. 10.2 Example of "brute force" mechanical shaker. Courtesy All American.

10.2 Reaction shakers develop considerably better motion. As phased in Fig. 10.3, vertical reaction forces add and left-right reaction forces cancel. The machine develops generally straightline motion, when the cg of the load is centrally located. Horizontal motion is also possible. There is some "hash" caused by gears, sliding and rolling bearing surfaces, etc.

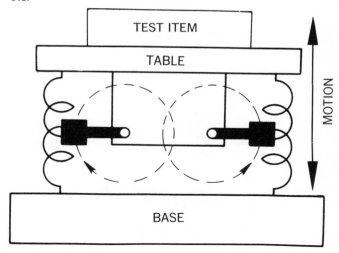

Fig. 10.3 Contra-rotating eccentric weights for softly-sprung "reaction type" mechanical shaker.

10.2.1 Frequency adjustments. Test frequency (in hertz, equal to unbalance mass RPM/60) varies with electric motor speed. A splined shaft and universal joints drive the table. Thus the drive motor does not vibrate. Frequency control has been programmed, but most machines are controlled manually.

10.2.2 Frequency limitations. Reaction-type machines are generally used only below 60 Hz, but a few have reached 120 Hz. Large machines with force capabilities to 300,000 pounds have been built, utilizing 150 horsepower drive motors. The waveform of motion is inferior to electrohydraulic (EH) shakers (Sec. 11) and particularly to electromagnetic (EM) shakers (Sec. 12). More importantly, EH and EM shakers can deliver random vibration.

10.2.3 Force adjustments. Force adjusting requires stopping the machine to adjust the angle between unbalance weights (Fig. 10.4), effectively changing the unbalance mass and centrifugal force.

10.2.4 Force limitations. The load rating is the greatest mass a machine can shake at a specified D and maximum frequency. There is also a static load limitation to be observed: the greatest load the springs can support. Machines capable of several inches stroke have been built.

10.3 Combined environment reliability tests or CERT tests, per MIL-STD-781B ("AGREE" tests at 2.2g RMS), originally used mechanical shakers, since frequency never exceeded 60 Hz. (However, the "C" revision in 1977, with its requirements for random vibration, led to many being replaced by EH and EM shakers.) The shaker vibrating table often serves as the floor of a climatic chamber.

10.4 Transportation tests often use mechanical shakers. See MIL-STD-810D on loose cargo transport. Typical frequencies of vertical or eccentric motion: 2 to 10 Hz. Typical accelerations: 1/2 to 2g, with drive motors to 50 hp. A typical goal: determining whether protective packaging is adequate for rail, truck, or air shipment. EH shakers capable of random vibration are rapidly replacing mechanical shakers.

Fig. 10.4 Adjustable eccentric weights. Courtesy L.A.B.

Fig. 10.5 Mechanical shaker combined with temperature chamber, as with MIL-STD-781B.

Section 11
Electrohydraulic Shakers

11.0 Introduction. For large forces, particularly with long strokes, as in lifting and positioning heavy loads, most engineers would consider hydraulic actuators. They can also be used for oscillating forces.

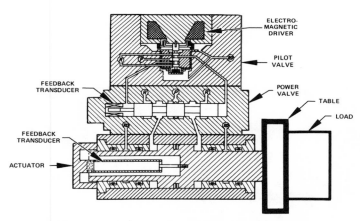

Fig. 11.1 Cross section of electrohydraulic (EH) shaker.

11.1 System familiarization. Fig. 11.1 suggests the internal construction of an EH shaker. Fig. 11.2 emphasizes the electrical controls. A servo valve modulates high-pressure oil flow from a pumping unit to the shaker. As oil is compressible, reasonably high f_fs require short oil passages. Thus the valve is mounted on the shaker. Commercial servo valves do not perform well above say 200 Hz; special high-frequency designs must be used.

Amplifier current in the electromagnetic (electrodynamic) driver results in vibratory force on the pilot valve. Valve motion uncovers valve ports. Oil flow causes power valve motion. This opens valve ports and (acting as a "hydraulic force amplifier") results in oil flow and actuator motion. Assuming a sinusoidal input signal, all parts should move sinusoidally. High frequency "dither" signals keep parts from sticking.

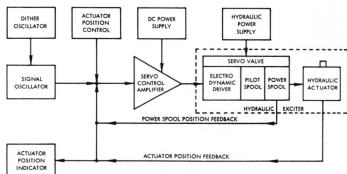

Fig. 11.2 Block diagram of EH shaker system for sinusoidal vibration. The system readily adapts to random control per Sec. 25 and 26.

Feedback transducers, usually LVDTs, sense position of the power valve and the actuator. These signals and appropriate analog or digital techniques instantaneously adjust piston motion so it closely follows the input signal. Looking at an oscilloscope, this is most obvious with sine inputs. Waveform is better than with mechanical shakers, though not as good as with EM shakers. $\ddot{x}$ waveform distortion (with a sine input) is claimed to be less than 10%. Distortion is also important with other motions including random vibration. Static force and position are separately controlled.

11.2 Limitations of EH shaker systems should be known to prospective users.

11.2.1 Displacement. Users can have long stroke or high frequency performance, but not both. Large D requires a long (and therefore compressible) oil column in the cylinder. Full 6 inch (152 mm) D on one machine can be attained only to 1 Hz, whereas full 1.5 inch (37 mm) on another can be attained only to 20 Hz.

11.2.2 Force and acceleration. The available force is basically equal to pressure × area. In the 6 inch (152 mm) D

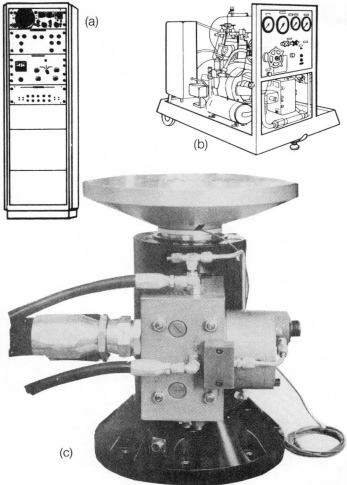

Fig. 11.3 Sketches of (a) controls and (b) motor-driven pump unit; photo (c) of small EH shaker. Note oil connections to servo valve. Courtesy MTS.

machine just mentioned, an effective pressure of 2,500 lb/in² (or 17 MPa) develops 71 klb (314 kN) force. Available A is calculated essentially as with EM shakers; see Sec. 12.3.

In one commercial system, 10 klb (44 kN) force is achieved with 1 inch (25 mm) D. The prime mover is a 125-hp electric motor which drives a 3,000 lb/in² (21 kPa), 70 gallon (265 liter)/minute pump. The full D can only be achieved up to about 3 Hz, but full force is available to about 400 Hz with one ton non-resonant loads. 10 inch (0.25 m) D is fairly common; some offer 18 inch (0.45 m) D. High-frequency performance is sacrificed. 21 foot (6.4 m) D is used for US Navy random vibration seasickness studies.

11.2.3 Velocity is also limited, by the hydraulic supply or by constrictions in the servo valve or elsewhere. Rather than be concerned about flow rates (gallons/min or liters/min), question the available inches/sec or mm/s velocity. The result is essentially the same as with EM shakers - see Sec. 12.6.

11.2.4 Frequency range. EH shakers readily develop a zero frequency (static) force. But what is the upper limitation? Authorities differ, partly depending upon their employers'

commercial bias. Those aiming at such markets as seismic, transportation, automotive and stress screening of large equipments suggest 200 Hz as an upper limit. Others, aiming at stress screening of small assemblies and at aerospace markets, claim 1,000 or 2,000 Hz. This is generally achieved with a deliberate oil-column resonance at say 1,900 Hz. See Isley's "Servo Hydraulics Vibration to 2,000 Hz," IES PROCEEDINGS, 1980.

The debate is perhaps academic. Even if 2,000 Hz piston vibration is achieved, can it be utilized? If you planned to shake many small units simultaneously, you would be interested in any table bending resonance. (All f_ns will drop with increasing load.) Significant acceleration gradients will exist across the table above the limits shown in Table 11.1. Constrained layer damping (see Sec. 1.3.4) will somewhat lessen this problem.

Table 11.1

Table size	f_n	Upper Limit
24 inches square	1,000 Hz	500 Hz
36 inches square	650 Hz	325 Hz
48 inches square	410 Hz	205 Hz
60 inches square	350 Hz	175 Hz

courtesy MTS

11.3 Random vibration with EH shakers. A major advantage of EH (and EM) shakers over mechanical shakers is the ability to create not only sine but also random and complex force and motion. See Sec. 25 and 26 for details of random control and Sec. 14 and 15 for details of sinusoidal control.

Fig. 11.4 Four EH shakers simulate road inputs to a pickup truck. Note shaker controls in foreground and random console (magnetic tape playback) in background. Courtesy Dana Corp.

11.4 Automotive/Truck applications. Groups of EH shakers can generate sinusoidal motion, either in- or out-of-phase, for engineering studies. Or (as in the road input simulation of

Fig. 11.4) each shaker can reproduce a motion that was FM tape recorded when driven on a test track. Alternately, random vibration can be synthesized. Complete vehicles as well as components (as in Fig. 11.5) may be tested much faster than driving on a test track. Such "ride simulators" are used by most automobile and truck manufacturers, worldwide.

Fig. 11.5 Fatigue test on portion of automobile suspension. Courtesy MTS.

11.5 Transportation tests on cargo effectively utilize EH shakers. Four, for example, drive the platform of Fig. 11.6. Tests identify weaknesses much more quickly than could extensive shipping, and failures can be observed as they occur. Also, different "fixes," or methods of strengthening or better protecting cargo can be quickly evaluated.

11.6 Earthquake simulation, utilizing EH shakers, is discussed in Sec. 28.4.

11.7 Multiaxis vibration, utilizing EH shakers, is discussed in Sec. 20.6.

11.8 Reaction mass. A 10,000-pound force EH shaker may weigh only about 125 pounds (mass about 57 kg). This permits easy positioning. However, for vibrating a massive load, the shaker must be rigidly connected to Earth or better to a reaction mass. Otherwise, the load will remain stationary while the shaker body destroys itself. A typical reaction mass will weigh about 10 times the shaker force, or, in this case about 50 tons. The mass usually "floats" on very soft springs (sometimes air springs), with an f_n of about half the lowest f_f; see Sec. 1.1.5.

11.9 Remote controls are particularly useful when EH shakers are used in hazardous areas. EH shakers have been attached to centrifuge arms for combining the stresses of steady-state and vibratory acceleration on components.

11.10 Torsional EH shakers are discussed in Sec. 29.4.

Fig. 11.6 Vibrating platform performs routine simulation of rail car transportation environment. Courtesy Ford Motor Co.

Section 12
Electromagnetic Shakers

12.0 Introduction — modal testing. This section briefly reviews the theory of electromagnetic (EM) or electrodynamic shakers, their development history and the wide variety available.

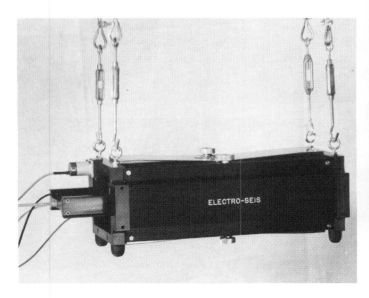

Fig. 12.1 Low force EM shaker, suspended for modal tests on relatively large structure. Courtesy Acoustic Power Systems.

Early (relatively low force) units were developed for modal studies, for experimentally determining how airplane structures respond to vibration, thus verifying designers' mathematical models. Not only are all flight vehicles modal tested, but nearly all engineering structures (as in Fig. 12.1) from computers to dams. Shakers are frequently used in pairs, such as one at each aircraft wingtip. Early tests were sinusoidal. As the f_f slowly varied, resonant effects were noted at f_ns. At each f_n, displacement and phase were read at many locations for modal mapping.

Fig. 12.2 A 36,000 lbf EM shaker. A relatively small load attaches to a relatively large table. Courtesy LDS.

Early sine testing was slow. Today's sine excitation is speeded by computer analysis and control. With random excitation, all resonant responses are excited simultaneously. Computers list the f_ns and can display and animate each structural response, per Sec. 17.2.1. Shakers aid in predicting responses of bridges to wind and vehicular loads; of automotive vehicles to road and engine inputs; of ship hulls to sea and propulsion system inputs; of nuclear power plants and their components to earthquakes, etc.

12.1 Theory of operation. Place a current-carrying wire in a magnetic field, Fig. 12.3; force will be generated in the wire. Coil the wire into a shaker driver coil (armature) in a radial magnetic field. Force produced will depend on number and length of turns, on magnetic flux density, and on coil current I. For vibratory force, use alternating current. The ratio between force (in RMS newtons or RMS pounds) and current I (in RMS amperes) is quite constant with frequency.

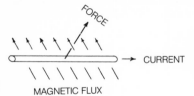

Fig. 12.3 The "right hand rule" which applies to motors, loudspeakers and shakers.

About 1954, the "single ended" design of Fig. 12.4 was popular. A typical table diameter was 0.4 m (16 inches), with armature mass 34 kg (weight of 75 pounds), developing a peak force of 15,400 N (3,500 lb). Magnetic shielding (not shown) reduced stray fields at the table to perhaps 100 gauss. Further reduction to 10 or 25 gauss was done with dc flow in "degaussing coils." This general design was used in a few shakers up to 66,000 N (15,000 lb) force.

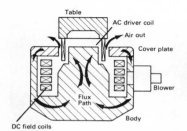

Fig. 12.4 Single-ended air cooled EM shaker design of the early 1950s.

To reduce stray field and gain higher flux concentrations for better efficiency (smaller amplifiers), today's "double ended" designs (Fig. 12.8 and 12.9) were introduced around 1959.

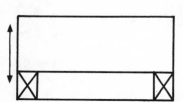

(a) Ideally, shaker armature motion is pure translation.

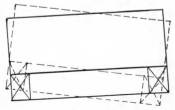

(b) Rocking armature motion often accompanies desired vertical translation.

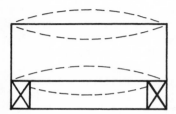

(c) A diaphragming or oil canning mode of table vibration often limits the useful frequency range.

(d) Axial resonance is shown here as alternate stretching and compression of the armature.

Fig. 12.5 Idealized shaker motion is often compromised.

12.2 Unwanted motions. Table motion is nominally perpendicular to the table surface, per Fig. 12.5(a). The effects here described occur during sine and random vibration tests, but are easier to detect with sine. Survey the surface with a hand-held "scanning" accelerometer per Fig. 5.33, or several fixed accelerometers. The motion will generally agree with (a) but at certain frequencies you will find (b) and (c). A phase meter or an oscilloscope connected to display "Lissajous" patterns (Sec. 18.9) aids in visualizing motions.

Table rocking is exaggerated in (b). Place two accelerometers at opposite edges of the table. At most frequencies the signals will be equal and in-phase; per (a). But at certain frequencies (due to unequal K among the flexure elements or non-symmetry of the load) the signals will differ in magnitude and phase. Rocking per (b) might "fail" an item where (a) would "pass." To quantize your test item inputs, one sensor at table center is not sufficient.

Table diaphragming (oil canning) is exaggerated in Fig. 12.5(c). Place one accelerometer at the table rim and another at the table center. At most frequencies the signals will be

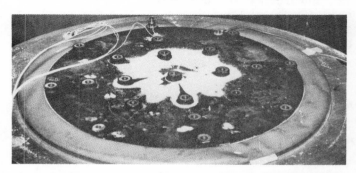

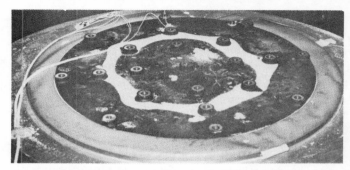

Fig. 12.6 EM shaker table resonant distortion demonstrated by salt patterns; 50g maintained at outermost attachment ring. Upper photo, 1,500 Hz, shows table center isolated. Lower photo, 1,800 Hz, shows modal ring.

equal and in-phase per (a). But at some frequency the table center will experience (worst at no load) up to 50 times the rim acceleration, per (c). With large-table EM shakers, this occurs under 2,000 Hz. One sensor at table center is not sufficient.

We mention (c) so you will appreciate that (1) no large structure (test item or test fixtures) moves as a unit (solid-body motion) at high f_fs; (2) large items often ignore (c); above say 200 Hz the internals of large (therefore soft) test items are isolated; much of the test is wasted.

Readers who have examined shaker armatures may question (c), because armatures *look* so rugged. Fig. 12.6 offers photographic evidence. These problems are most troublesome with large diameter tables. Head expanders accommodate large loads (see Fig. 20.3), exaggerating this problem.

Axial resonance is exaggerated in Fig. 12.5(d). It little affects test quality, but if sustained will damage the shaker armature. Test specifications should limit motions (b) and (c).

In the marketing department, however, another force rating is added: 62,000 N (14,000 pounds) peak sine force, based on the 1.414 peak/RMS ratio (which *only* applies to sinusoids). In F = MA calculations, A represents peak sine acceleration.

Available A is calculated as follows:

"USA" SYSTEM

$$A = \frac{F}{W} \qquad (12.1)$$

where A is in peak "g" units,
  F is in peak pounds force,
  W is total moving weight, lb.

INTERNATIONAL SYSTEM

$$A = \frac{F}{M} \qquad (12.2)$$

where A is in peak m/s²,
  F is in peak newtons,
  M is total moving mass, kg.

Fig. 12.7 Extending table diameter provides additional table area. Courtesy Thermotron.

12.3 Heat limits F = MA ratings of EM shakers. Temperature rise (and differential expansion of parts) in the electrical windings limit the available force. Engineering and production personnel within an EM shaker firm might RMS rate a shaker (assuming an adequate PA) at 44,000 N (10,000 pounds). This permits F = MA calculations per Sec. 23.10; A, the standard deviation σ is given in RMS units for random vibration tests. Momentary peak forces (for random vibration tests with this shaker) can reach 132,000 N (30,000 pounds).

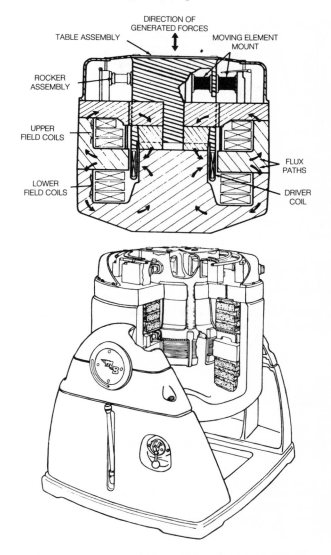

Fig. 12.8 Double-ended EM design with cooling provided by immersion in circulating oil. Courtesy MBIS.

12.4 Cooling. The windings in Fig. 12.8 are cooled by oil flow. Those of Fig. 12.9 are wound of hollow conductors that carry current and distilled water. Both depend upon an external heat exchanger. Such liquid cooling is quieter (operator better hears impending shaker or test article failure), and heated air is not dumped into the laboratory. For a given frame size, liquid cooling permits about double force *if* four times the power is available. Since armature M is not increased, light load A can be doubled. Thus 100g or greater A is possible.

Fig. 12.10 Armature for 40,000 lb. RMS double-ended EM design. Mass is 111 kg. (weight 245 lb). Input power is about 230 kva. Courtesy U-D.

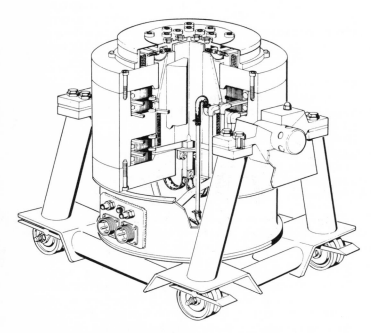

Fig. 12.9 Double-ended EM design utilizing coolant flow through hollow conductors. Courtesy Ling.

12.5 Armature. Whereas the body (magnetic circuit) is steel, the table (armature) is often cast of a magnesium alloy (see Fig. 19.7) having low mass density and good damping. Low armature mass permits heavier loads or higher g levels. Stainless steel inserts in the soft table permit bolting test specimens. Copper or aluminum wire or tubing is used for the windings.

See Fig. 12.10. The outer bolt circle diameter is 0.6m or 24 inches; the reader can estimate other dimensions. First resonance is slightly above 2,000 Hz. See also Sec. 18.6.

The design of Fig. 12.11 eliminates the greatest weakness of conventional shaker designs: the flexible ac (and any coolant) connections to the moving coil.

12.6 Velocity limits. Shaker stroke limits per brochures and instruction books (typically 25 or 50 mm, 1 or 2 inches D), are fairly obvious. But full D is only available at low f_s, per Fig. 12.12. A increases per $.00202f^2D$ (mm) or $.0511f^2D$ (inches). The full stroke "constant D" line can only rise to the A limited by F/M. With light loads, A and D are further limited by V. V limits are always given (but seldom understood) in brochures and instruction books.

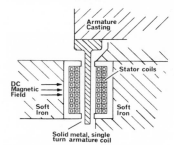

Fig. 12.11 Section through single-turn aluminum plate driver coil. Note primary windings (in pole faces) which induce thousands of RMS amperes into moving secondary. Courtesy U-D.

Visualize the driver coil moving in an intense magnetic field. In that coil is induced an EMF proportional to winding length, to flux density and to V. This EMF opposes the amplifier voltage forcing current through the coil. If the amplifier develops enough voltage, no problem. But an amplifier of inadequate voltage (perhaps purchased some years earlier when D and A were smaller) cannot fully drive the shaker. With EH shakers inadequate oil flow similarly limits V.

Maximum sine test V occurs at f_c, per Sec. 14.2. Before scheduling a test, calculate V. If too high, get another shaker and/or amplifier. Exceeding the V limit produces "clipping" and severe distortion. Your ear (and your oscilloscope) will alert you to STOP THE TEST!!!

A possible further V limit: heat generated in the flexures (see Sec. 12.7). Coolant should be directed to each flexure.

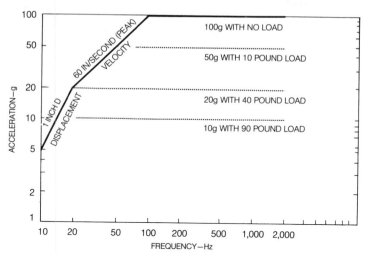

Fig. 12.12 *Typical limits of shaker D, V and A for various mass loads.*

12.7 Flexures suggested by Fig. 12.13 guide the armature, limiting lateral and rocking motions in spite of load eccentricities. A variety of flexure styles is used to permit one or two inches D, including ellipses and half ellipses made of alternating layers of metal and elastomer (for damping), rolling struts, cantilever beams, bonded rubber to metal shear mounts, air bearings, hydrostatic bearings and recirculating ball bushings. Sec. 12.2 suggests instrumentation for warning of flexure deterioration. Straight-line motion is most important in calibrators per Sec. 9.9, less important for general-purpose vibration testing. Indeed, two and three axis motions are increasingly provided per Sec. 20.6. See also the fixture requirements of Table 19.1.

Fig. 12.13 *Cantilevers, one of many flexure styles. Courtesy Thermotron.*

12.8 Variety of shakers. Over the past 35 years, a rather bewildering array of EM shakers has been built, with many still in use. Details have changed, but not physical principles. Sec. 12 illustrations show the variety.

12.9 Maintenance ease should be a design goal. Armatures, flexures, fluid and electrical connections to driver coil, etc, can fail. They receive the same damaging vibration, day after day, as do test parts. The armature assembly and field windings should be easily removable.

Every Monday, make an empty table "waveform check". At full rated force (full D or V at low frequencies), sine input, slowly sweep the frequency range. Monitor (with an RTA or at least a 'scope) an accelerometer mounted on the table. Record the amount of any distortion at various f_fs. If distortion has increased, seek and remedy the cause (***before*** failure occurs) such as looseness of, or a crack in, some part of the shaker. Or something rubbing. Otherwise, a catastrophic and expensive failure may occur, usually during "the most important test of the year." Replace any worn parts. Check all connections. Consult service notes for maintenance.

Also measure current needed for full rated output at various f_fs. Changes may indicate deterioration in the shaker, the power amplifier, and/or the measurement system.

12.10 Spare parts should be available so you can quickly correct a shaker failure and resume testing. Stock a spare moving-coil assembly (possibly a complete armature), spare flexures, etc. If a shaker is engaged in a "crash" program or is far from a parts depot, spares are worth their cost.

Consider rebuilding the original armature, as suggested by Fig. 12.14. Then the rebuilt armature becomes a spare, ready for the next emergency.

12.11 Combined environments (temperature, humidity, vibration and altitude) are demanded by Method 520.0 of MIL-STD-810D to "identify failures .... can induce in aircraft electronic equipment .... during ground and flight operations." Interactions between these environments can produce a variety of equipment failures.

Fig. 12.14 *Checking Ling shaker armature after rebuilding. Courtesy ACG. Rebuilding firms may offer used shaker systems.*

Liquid cooling was first used with shakers inside chambers for "combined environment" (altitude, temperature and humidity extremes) testing. Air cooling could not be used. Most shakers today are used outside chambers, with "table extenders" protruding through chamber floor or walls per Fig. 12.15.

Fig. 12.15 EM shaker driving through chamber floor for combined environment testing. Courtesy LDS.

12.12 Shaker and centrifuge. A very few users combine shaking with unidirectional forcing by placing a modified shaker on a centrifuge arm.

12.13 Multiple shakers for large loads. Fig. 20.14 suggests multiple EM shakers used with loads which dwarf the shakers. Several EM shakers can be driven by a single amplifier/control system. More commonly each shaker has its own power amplifier. For sine testing, one control unit (see Sec. 14 and 15) usually acts as "master" with the other controls "slaved." Controlling the relative phase angle between them requires complex phase-controlling systems; none are perfect. For multi-axis random tests, each shaker usually has its own noise generator and equalization (see Sec. 24-26). For single-axis random tests, all shaker signals come from a single noise generator.

Section 13
Power Amplifiers for Driving Electromagnetic Shakers

13.0 Introduction — what does the power amplifier do? It accepts a small electrical signal from an oscillator (Sec. 14.1) or a computer (Sec.15) for sine testing. Or from magnetic tape (Sec. 24.1). Or from a noise generator (Sec. 24.2) or a computer (Sec. 26) for synthesized random vibration. The power amplifier (PA) makes that signal larger, delivering from a few hundred watts to 100 kw or more, depending on EM shaker size.

13.1 EM shaker needs: enough dc power to excite the shaker's magnetic field, also enough ac power to drive the shaker's moving coil (see Sec. 12.1).

13.1.1 DC power. Fields of small shakers (up to 440N or 100 lb. force) typically use permanent magnets. Medium size shakers develop dc from single phase mains. For larger shakers, three phase power is rectified; here ripple, which can cause unwanted shaker motion (especially during low force tests) is easier to filter. "Switched" dc supplies rectify at still higher frequencies (above 20 kHz), for easiest filtering and least heat losses. Modern EM shakers use much cheap dc power to conserve expensive ac power from the PA.

13.1.2 AC power is needed for the driver coil. Fig. 13.1(a) suggests the current necessary to generate constant "no load" motion at various f_fs. The low f_f notch is due to mechanical resonance of the mass (table + load) on the flexure springs. The high f_f notch is caused by armature

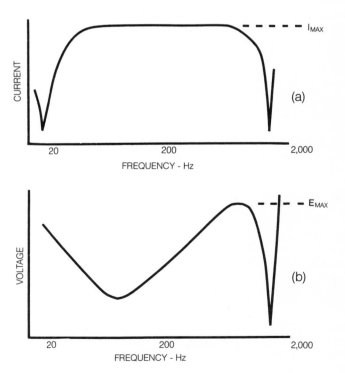

Fig. 13.1 EM shaker ac current and voltage requirements.

mechanical resonance per Sec. 12.2. Fig. 13.1(b) suggests the voltage needed to force this current through the coil. These graphs change markedly when loads, especially resonant loads, are added. Sec. 14.6 deals with electronic compensation for Fig. 13.1(b), enabling control of test load input motion.

The PA must deliver enough voltage E and current I to reach the maxima of Fig. 13.1. E and I are both RMS quantities; they establish the needed kVA rating, into a highly reactive (0.2 capacitive to 0.2 inductive) load. The largest rating known to us (1984) is 256 kva. Many random vibration tests call for brief severe $\ddot{x}$ peaks per Sec. 22.10. E and I peaks must briefly exceed their RMS ratings.

13.1.3 Comparison with hi-fi amplifier. Shaker amplifiers resemble high-fidelity, wide f_f, high continuous power sound system amplifiers, with the ability to reproduce high peaks. However, shaker users critically *measure* system performance (using accelerometers and distortion analyzers), so distortion must be very low.

13.2 Still some tube-type amplifiers in use. Until about 1955, shakers were driven by motor alternator sets. Then came PAs featuring vacuum tubes. By 1980 all USA manufacturers had "gone solid-state." In 1984 a few older units still use tubes; difficulty in obtaining spares (Sec. 13.10.5) will soon retire these.

13.3 Advantages of solid-state amplifiers: (1) typical supply voltages are ±75, much safer for maintenance than +5,000 to +15,000 or higher; (2) less danger of breakdowns in rectifiers, filter capacitors, transformers and wiring, so higher reliability; (3) full power almost instantly available; (4) less heat generated, so (5) cooling requirements are eased and (6) less electric power consumed; also (7) greatly improved low-f_f shaker performance, as an impedance-matching output transformer is not generally needed. Full shaker performance to essentially zero f_f is possible, helpful in testing at say 25 mm (1 inch) D at 2 Hz, also for some shock testing on shakers. Up to four PAs have been ganged to a single shaker, generally for shock testing (Sec. 33.4.4).

13.4 Cooling. Smaller amplifiers are air cooled; larger rated (but relatively compact) units are quieter as cooled by distilled water through the output (and sometimes driver) stages.

13.5 Operation. Power amplifier design must insure

> safety for operating and maintenance personnel,
> easy and logical operation
> maximum reliability, and
> ease of service and repair.

Operation is here discussed in very general terms. Turning on an older high voltage tube-type amplifier usually involves (1) Turn on amplifier and shaker cooling. (2) Turn on single phase control circuits: oscillator, servos, random noise generators, computer, electronic meters, etc. (3) Verify all interlocks (doors, cooling, protective system resets, and others) are closed. Turn master gain control to zero. Then (4) turn on main power. Some systems progress automatically through delays involving (4a) heater power to tubes, (4b) shaker field energized, (4c) grid bias supply on, (4d) screen

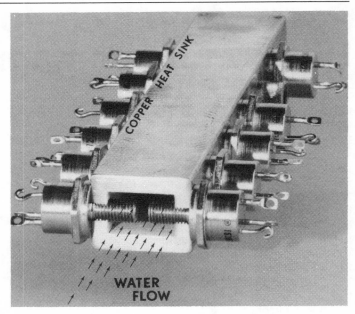

Fig. 13.2 Power transistors in water-cooled heat sink. Courtesy U-D.

supply on and finally (4e) plate supply on. Other systems require operator switching. Increase the master gain control to start the shaker vibrating.

Fig. 13.3 2 kw power module, has been ganged to 192 kw. Courtesy LDS.

In comparison, turning on a solid-state amplifier is simple. No warm-up is required, and full shaker force is immediately available on (4).

Fig.13.4 32 kw power amplifier, 16 modules per Fig. 13.3. Courtesy LDS.

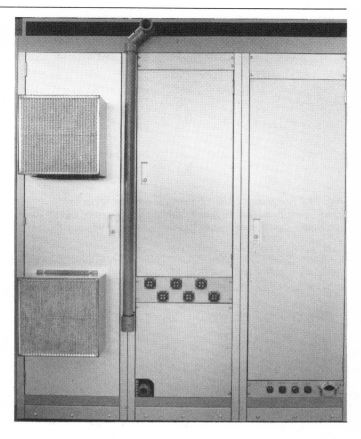

Fig. 13.5 Large power amplifier. Fixed installation usually requires riggers, plumbers and electricians. Courtesy U-D.

Operating controls and signal sources should be near the shaker. If controls are in a separate room, the operator must see (window or TV system) and hear (headphones or loudspeaker connected to the control accelerometer) the shaker to detect any malfunction in shaker or test article. He needs an intercom system to communicate with men at the shaker.

13.6 Importance of low distortion. As shown in Fig. 13.1, little current and voltage are needed for full motion at armature electrical resonance 100 to 200 Hz per Fig. 13.1 and Sec. 14.4.1. Here the shaker is very "efficient."

Don't be surprised when shaker motion, as viewed on an oscilloscope, is not sinusoidal. It often resembles the waveforms of Fig. 13.6, particularly the third harmonic waveforms. What causes this distortion? Two factors must be present simultaneously:

1. A high-Q (lightly damped) mechanical resonance in the shaker armature, the test fixture or the item being tested. An armature resonance, with a Q of say 100, may occur at say 3,000 Hz, as in Fig. 13.7.

2. Amplifier harmonic distortion at that frequency. (Third harmonic distortion is usually the most troublesome.)

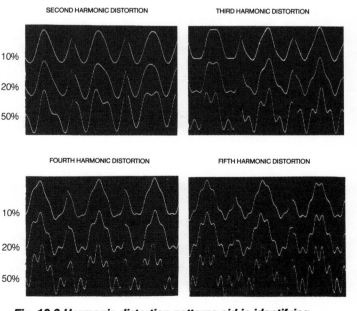

Fig. 13.6 Harmonic distortion patterns aid in identifying shaker waveforms. Each top row represents 10%, center row 20% and bottom row 50% distortion. Columns represent different phase angles between harmonic and fundamental. Memorize these patterns for evaluating shaker behavior. Or use a distortion or spectrum analyzer. Courtesy Chadwick-Helmuth.

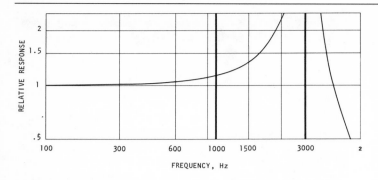

Fig. 13.7 Response of SDoF system, emphasizing below-resonance behavior. Text uses this curve to model EM shaker response. Response values increase force at various f_fs.

Imagine testing at 1,000 Hz; current to the shaker (and thus its force) is 99 parts 1,000 Hz and 1 part 3,000 Hz (1% distortion). Estimate the 1,000 Hz motion

$$99 \text{ parts force} \times \text{response factor } 1.09 \times \frac{\text{force}}{\text{mass}} = 108 \text{ parts.}$$

Estimate the 3,000 Hz motion

$$1 \text{ part force} \times \text{response factor } 100 \times \frac{\text{force}}{\text{mass}} = 100 \text{ parts.}$$

Thus when we test at 1,000 Hz, we have about equal motion at both f_fs, 1,000 and 3,000 Hz. Every time f_f passes through 1,500, 1000 especially, 750, 600, 500, etc., some 3,000 Hz. vibration will occur. At these f_fs factors 1 and 2 combine; also at submultiples of other shaker resonances, of fixture resonances, and of test item resonances. Should our test item fail, we would be unsure of *which* f_f caused failure.

Harmonic distortion also affects automatic servo controls, unless a "tracking filter" (Sec. 14.7) is used. Worse, motion at 3,000 Hz may exceed the specified test (f_u typically 2,000 Hz). Isn't this an overtest? Trusting $f_u = 2,000$ Hz, a designer might deliberately set an f_n at 3,000 Hz. That item may wrongly fail when $f_f = 1,000$ Hz.

A question for Quality Control: how much distortion will you permit? Test standards are amazingly silent on this important point. Only one official document mentions it: NAVORD OD 45491 (available from Naval Ordnance Station, Louisville, KY 40214); its Sec. 2.8.1.3 limits shaker distortion to 10% of the fundamental. At least one shaker system supplier will guarantee 5%, except at 1/2, 1/3, etc. of f_n.

Shaker amplifiers thus demand extremely low distortion, difficult to achieve with a reactive load. Our example used 1% distortion. (For comparison, many hi-fi PAs have 3% or more distortion.) Push-pull design cancels even harmonics, leaving mainly 3rd, and to a lesser extent, 5th, 7th, etc. harmonics.

Negative feedback improves performance per Fig. 13.8. The broken curves show the flat response and zero phase shift frequency ranges widened, to amplify complex and random signals with least distortion. Also, negative feedback reduces output impedance; shaker and load resonances less distort motion.

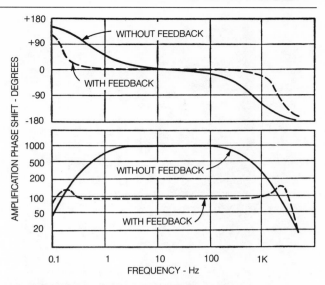

Fig. 13.8 Amplifier phase and magnitude response.

Distortion rises quite rapidly as full PA power is approached, per Fig. 13.9. We recommend 2:1 or more reserve capacity. Do you recognize the similarity to hi-fi music reproduction?

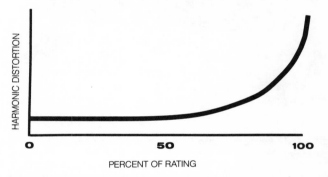

Fig. 13.9 Harmonic distortion rises as full PA rating is approached.

When you evaluate a shaker system, either for maintenance, when accepting a new system, or when considering purchase, slowly sweep all f_fs for evidence of harmonic distortion in the $\ddot{x}$ waveform.

13.7 What do amplifier ratings mean? Shaker system proposals include a PA description, meaning little to most prospective purchasers. Consider, for example, the amplifier manufacturer's numbers for harmonic distortion (as little as 0.05% under idealized but impractical conditions), also for background (no signal) hum and noise. His measurements refer to a resistive load; but since shakers are highly reactive, these numbers have no value. Ask the system manufacturer to state SYSTEM performance to include not only D, V and A limits but also the amount of harmonic (and other) $\ddot{x}$ distortion *at the shaker*.

13.7.1 Importance to random tests. The claim has been made that PA distortion is meaningless in random vibration testing. True, a 'scope time history per say Fig. 22.4 does not show PA distortion as easily as does Fig. 13.6. However, a spectrum analyzer can show unwanted 3 to 10 kHz vibration (when f_u is *supposed* to be 2 kHz).

Consider 3σ clipping (Sec. 27.4), for example. Your authors remember when PAs were always designed for E and I peaks of 3σ. But now we see PA input circuits designed to clip at say 2σ; this permits increasing the shaker system RMS rating. Per Sec. 27.4, this creates unwanted shaker motions 3-10 kHz. This is "legal," unfortunately; few MIL standards mention the quality of shaker motion.

13.7.2 Hum and noise measurements, checked with zero PA input, should not exceed 0.1g "background" shaker motion. Even this is troublesome during a resonance investigation at say 1g. 0.02g *can* be achieved; why not require it?

13.7.3 Harmonic distortion checks. Check shaker motion distortion at maximum shaker force (limited by D, V and A) over all f_rs with sine input. Most EM shakers have one or more high-Q peaks in the range 1-3 kHz and amplifier-caused distortion will be greatest at 1/3, 1/5, etc. of those f_rs. Don't be fooled by clean "bare table" checks; if shaker armature Q is low, amplifier distortion may seem negligible. But tests on high-Q specimens and high-Q fixtures can be difficult. A better check: mount a flat plate with corners overhanging. With fingertips, ears or an accelerometer on a corner, identify the corner bending f_n. Check distortion at 1/3, 1/5, etc. of that frequency.

13.7.4 Visit factory. Put system performance limits into your requests (to shaker system manufacturers) for proposal (RFP). Tell all prospective suppliers (via the RFP) what is important to your laboratory. Mention D, V and F requirements. Add limits such as 0.02g for hum and noise, and 10% for shaker distortion. State how these are to be checked.

State that you will visit the factory and verify all system ratings before shipment. This will cause some anguish. Shaker manufacturers prefer to ship systems and let field personnel deal with problems. But field personnel are limited in what they can accomplish. Also, once the system arrives in your lab, you will be hurried to accept the system (in spite of difficulties) and to commence long overdue tests. Far better to keep the system at the supplier's plant, where engineering and production people can work together.

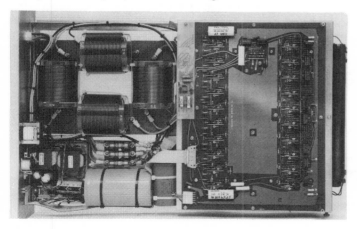

Fig. 13.10 Power amplifier module switches at 56 kHz for small size and high efficiency. Courtesy MBIS.

Request installation information (for your plant engineering people) + all system and unit wiring diagrams + all instruction and maintenance manuals, all to be shipped two weeks ahead of the system. Study the diagrams. Read the manuals. Prepare a list of questions to ask supplier installation personnel. This will speed "putting the new system to work."

13.8 Amplifier efficiency was of little concern until recently. Excess heat was dumped. With today's increasing energy costs, plant engineers attempt to recover that heat. Even better, users seek higher PA efficiency, wasting less power. For example, a Class D "chopped" or "switched" amplifier that delivers 36 kva needs 45 kva input, whereas conventional Class AB units may require 120 kva input. Note that some AB units also offer economies in "idling input power," since full output power is only needed briefly.

Secondary benefits of switching: less space and less cooling (generally air instead of water) needed. Problems include electromagnetic interference (EMI). Power supply rectifiers (Class D *or* AB) create EMI. So does switching above 20 kHz. Special enclosures are needed, with special filtering on all cables. Shaker hum and noise may be higher than with conventional AB designs. Be sure to check low level $\ddot{x}$ waveforms at say 10 Hz; waveforms may have troublesome spikes.

Evaluating "switched" vs. conventional AB amplifiers is complicated by disagreements between amplifier "experts" at technical meetings, in published papers and in sales brochures. Few firms build both kinds; each supplier defends his own position. Protecting "switched" amplifiers against overload is said to be difficult.

13.9 Which should we buy? (Much of this section applies to shakers and controls, as well as PAs.) Very likely a supplier can demonstrate higher efficiency or lower shaker distortion with either kind of amplifier, given circumstances favorable to his position. We recommend that you somewhat discount sellers' specifications. Discount heavily *your* ability to write a "tight" specification. You are less experienced than the suppliers.

Base your decision, rather, on long-term factors important to your laboratory's success. Visit other laboratories and talk, at technical meetings (see Sec. 35.3), to other shaker users. Ask questions about the various suppliers, about their ability and helpfulness AFTER the sale. Will Supplier X still be in business in 5 years?

13.10 Power amplifier maintenance, in leading test labs, is performed by the operators. They observe symptoms during tests, then diagnose difficulties. We deplore union restrictions and job classifications which limit operators to attaching and removing test items while a computer runs tests. They must not fix a shaker, but rather must wait for maintenance personnel (hopefully trained on shaker systems) to arrive. Such laboratories have many difficulties. Better to train operators to maintain their equipment. One example: an alert operator will hear such changes as a broken flexure.

13.10.1 Electrical safety cannot be overemphasized. Be very careful when entering or reaching into any tube-type amplifier. Voltages to 15,000 can be fatal. Solid-state amplifers have different dangers: rings, tie clasps, etc. can short circuit high current supplies and cause bad burns. Don't cheat protective devices.

13.10.2 System operational checks. Maintenance manuals describe operational checks as well as troubleshooting routines. Here we discuss a weekly "wellness" check of an EM shaker system. Per Sec. 13.7.2, with the master gain control at zero, measure and record "zero signal" table motion. Has it increased? Analyze the spectrum. At what f_f was the increase? This may suggest needed maintenance.

Operate the shaker at its full rated output (full D, V and A) through the entire f_f range. At critical f_fs, such as 1/3, 1/5, etc. of a high-Q resonance, measure and record the applied voltage (across the shaker driver coil — you may have to add such a meter), also the $\ddot{x}$ total harmonic distortion. Has it increased? Analyze the $\ddot{x}$ spectrum. At what f_f was the increase? This may suggest needed maintenance.

Such checks often expose difficulties in the shaker, the PA, controls and interconnects. These can be diagnosed and remedied much better on Monday than on Friday during an important test.

13.10.3 Regular inspection needed. Visually inspect the PA (power off) weekly, oftener with heavy use and with bad environmental conditions. Carefully remove dust buildup from forced-air cooling. Check terminal strips; tighten any loose connections, possibly caused by vibration. Clean any dirty contacts or exposed switches and relays. Replace any pitted or burned contacts. Clean up any fluids due to leaks, spills or condensation.

13.10.4 Keep a log of all malfunctions, showing dates of maintenance checks and any component replacements, circuit modifications, etc.

13.10.5 Spare parts stocked depend upon many factors: (1) Effect of a system shut-down: if you have only one shaker system, and if tests *must* stay on schedule, stock many parts. (2) if you are far from sources, stock many parts. (3) Time to get a part, including delays in your purchasing department, supplier delays, air freight schedules, etc. (4) Financial ability to tie up thousands of dollars in spare parts. You can temporarily replace many parts such as resistors and capacitors with electrical equivalents; don't stock these. (5) Possibly a nearby lab has a similar shaker system and can share the parts stocking burden.

13.10.6 Troubleshooting. If your system malfunctions, try to determine which major unit (PA, shaker, or console) is causing difficulty. Don't blame your PA for difficulties elsewhere. For example: a system might seem underpowered at 1,000 Hz and above; the actual fault might be a too-soft test fixture.

Solutions to difficulties should be in the manufacturer's instruction and maintenance manuals. If not, telephone his service people; they often diagnose difficulties by phone. Or they can send field service people. Should all these approaches fail, write or call us.

Suppose that on Monday you have run your system at full rated F over its entire f_f range, have observed table $\ddot{x}$, and have observed the same driver coil E and I as in previous weeks. Now, if trouble appears during a test, question the specimen and fixture first, rather than waste time on the shaker and PA.

Unplanned PA shut-downs or "dumping" do not necessarily mean PA trouble; your shaker may have failed or an interlock may be triggered. Complex systems provide panel indicators (as in Fig. 13.11) to show where trouble exists.

Fig. 13.11 Diagnostic PA control panel, usually located in control console. Courtesy U-D.

If you suspect PA trouble, try to localize the problem. Suppose there is no shaker motion, yet power is on. Use a voltmeter or 'scope to verify arrival of an input signal. Shaker driver coil meters indicate zero ac voltage and current. There is no signal into the final amplifier. Therefore the trouble must be in a preamplifier or its power supply; check their test jacks (first disconnect power to the final amplifier stages).

You may wish to check the PA (distortion, for example) with a resistive load rather than the reactive shaker load. Place electrodes in a barrel of salt water.

Logical, clear thinking, and not "jumping to wrong conclusions" will speed locating and repair of faults. This can be difficult under "customer" pressure to resume testing, another argument for preventive maintenance. Operators need much training and study. They must become familiar with systems so that (1) fewer breakdowns will occur, and (2) in the event of trouble, service will be restored promptly.

We greatly appreciate the assistance of Andrew Emerson, Harry Hosford, Fred Hunt and Robert Turner in preparing Section 13.

Section 14
Analog Controls
for Sinusoidal Testing

14.0 Introduction. Sec. 14 and 15 apply to both electro-magnetic (EM) and electrohydraulic (EH) shaker systems. EM shakers will be used for most examples, as they are more common.

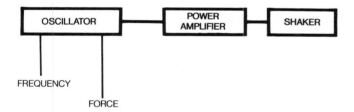

Fig. 14.1 Basic elements of manually controlled sinusoidal vibration testing system.

Fig. 14.2 is more complete than Fig. 14.1; a test load is attached to the shaker. A "control" accelerometer on the shaker, connected to a readout meter, tells the operator how intensely his shaker is vibrating. Under manual control, he controls f_f with his left hand. He controls vibration intensity with his right hand.

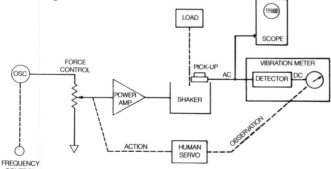

Fig. 14.2 Similar to Fig. 14.1 but further detailed. Note that operator closes the loop to maintain specified vibration intensity.

14.1 Oscillator. This unit provides a sinusoidal signal at f_f. A typical range is 10 to 2,000 Hz. Its signal should be pure; the power amplifier output can be no cleaner than its input. If the signal contains distortion (additional frequencies), test results may be misinterpreted. With f_f at 1,000 Hz, and appreciable say 3,000 Hz third harmonic, a test item resonance might be wrongly identified as a 1,000 Hz critical frequency, when the actual resonance is at 3,000 Hz, outside the test range and therefore to be ignored. (Same effect as power amplifier distortion.) Typical new oscillators specify 1/2% total distortion; this often worsens with time. 0.1% distortion is desirable, but seldom achieved. A servo unit per Sec. 14.6 may further distort.

1950s RC and beat frequency oscillators displayed frequency on a dial, with say ±1% accuracy. Better: an electronic counter to display frequency. Test item critical frequencies must be recorded very accurately in test reports.

Fig. 14.3 Voltage-controlled audio oscillator suitable for shaker control. Separate unit on top provides multiple levels per Section 14.3 and Fig. 14.8. Courtesy Scientific-Atlanta.

Later voltage-controlled oscillators (VCOs) are tuned electronically. The unit of Fig. 14.3 provides either linear or logarithmic (see Sec. 14.1.2) sweeping of f_f. For manually identifying resonances, the operator relies upon "peaking" a meter (connected to a "response" accelerometer somewhere on his test item) to indicate that $f_f = f_n$.

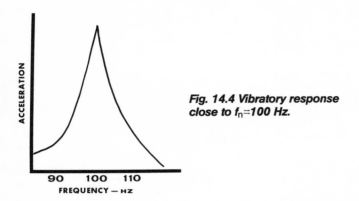

Fig. 14.4 Vibratory response close to f_n=100 Hz.

14.1.1 Automatic sweeping needed.
Manual f_f control is fine for some tests, but not for repetitive sweeping. Visualize an operator whose specimen has a critical frequency, say at 100 Hz, per Fig. 14.4. Some damage occurs during each up and each down sweep, during the few seconds of resonance close to 100 Hz. His attention wanders. Suppose he slows or stops at 100 Hz. 100 damaging cycles occur every second until he resumes sweeping. The specimen fails. Will anyone blame him? Probably not. If the specimen is a prototype, the failure may wrongly "lose" a contract or cause redesign. Programmed frequency control is needed.

14.1.2 Logarithmic vs. linear sweeping.
Early standards vaguely specifying "uniform" frequency variations were interpreted in at least two ways per Fig. 14.5.

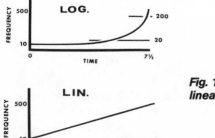

Fig. 14.5 Logarithmic vs. linear frequency sweeping.

Imagine sweeping 10-500-10 Hz. A test item part resonates at 20 Hz; prolonged 20 Hz testing is damaging. To "pass" the test, should you choose linear or logarithmic sweeping? To most quickly traverse the damaging 20 Hz region, specify linear sweeping; note which graph is steeper at 20 Hz. However, to prolong 20 Hz resonance stressing, choose logarithmic sweeping; it more slowly traverses this region, and more quickly traverses say the 200 Hz region.

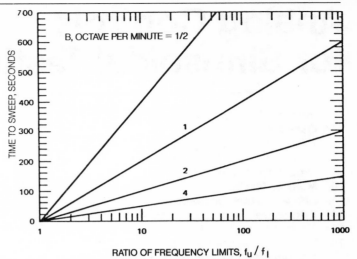

Fig. 14.6 Nomograph for duration of logarithmic frequency sweep.

Most test standards require logarithmic sweeping. If not explicitly stated, watch for such a key phrase as, "___ minutes per octave (or per decade)." Sweep time in seconds relates to sweep rate B (octaves per minute), to f_u and to f_l by eq. 14.1 and by Fig. 14.6.

$$\text{Time} = \frac{60}{B \ln 2} \ln\left(\frac{f_u}{f_l}\right) (\text{sec}) \qquad (14.1)$$

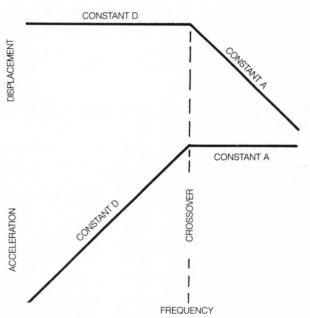

Fig. 14.7 Typical swept-sine test program shown (upper) as D vs. f and (lower) as A vs. f.

14.2 Crossover and crossover frequency.
Many standards (Fig. 14.7 provides an example) call for two vibration intensity control modes:

   1. Constant D at low f_fs; a monitoring accelerometer signal is twice integrated to control D.

2. Constant A at high f_fs; a monitoring accelerometer signal is used directly to control A.

The controller automatically changes mode at the "crossover frequency," f_c. Here the specified D and A occur simultaneously. Calculate f_c by transposing eq. 2.10 (if D is in m and A is in m/s^2) or by transposing eq. 2.11 (if D is in mm and A is in g units) or by transposing eq. 2.14 (if D is in inches and A is in g units), and then solving for f.

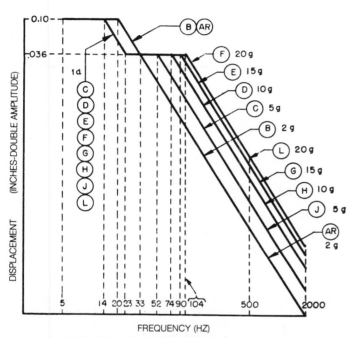

Fig. 14.8 Test personnel can select various standard A values, per MIL-STD-810C.

14.3 Multiple levels as in Fig. 14.8 are required by some Standards. They require two or more "crossovers" from D to A and vice versa. The controllers of Figs. 14.3 and 14.16 provide such flexibility.

14.4 Shaker frequency response. If we vary f_f, shaker table A will vary per Fig. 14.9. As compared to a constant A, we will overtest at frequencies f_2 and f_4, and undertest at frequencies f_1 and f_3.

14.4.1 Armature electrical resonance. Why? What causes shaker "efficiency" to vary per Fig. 14.9 (necessitating corrective operator action per Fig. 14.10)? We will use electromechanical analogs; shaker and load mechanical properties are represented by analogous electrical characteristics. These, along with shaker electrical properties, explain shaker frequency responses without mathematics.

In Fig. 14.11 and beyond, we apply a swept sine constant voltage to the driver coil and measure the resulting shaker table A. With more mass load, A decreases per F/M as in (a). Also, note that the frequency of peak "efficiency" drops. (b) represents the armature mass (and any attached mass load) by a capacitor C; it behaves electrically as does a mass mechanically.

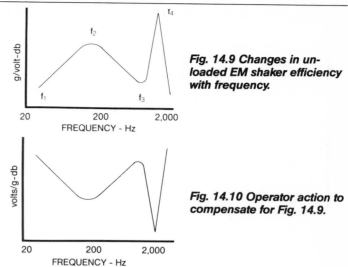

Fig. 14.9 Changes in unloaded EM shaker efficiency with frequency.

Fig. 14.10 Operator action to compensate for Fig. 14.9.

At some frequency in the range 50 to 200 Hz the driver coil electrical inductance L_{coil} and C will resonate. At this quite broad resonance (often called electrical resonance), least voltage is required per g. Do not confuse this resonance (f_2 in Fig. 14.9) with the mechanical resonance (at perhaps 15 or 20 Hz) of the armature plus load, suspended upon spring-like flexures.

Fig. 14.11 (a) Changes in electrical resonance as mass load is increased from zero. (b) Electrical analog; C represents total mass (table + coil); L_{coil} represents inductance of driver coil; R_{coil} represents resistance of driver coil; E_{coil} represents applied voltage; E_{out} represents table motion (velocity).

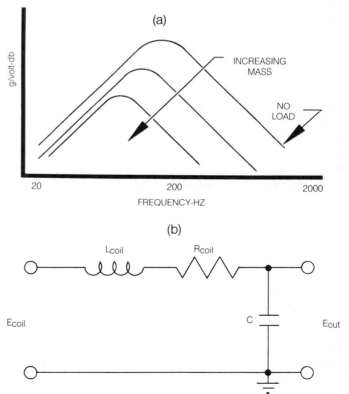

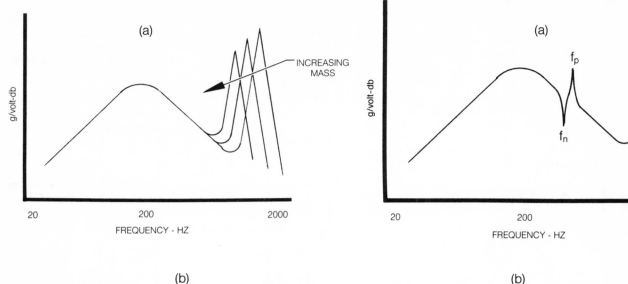

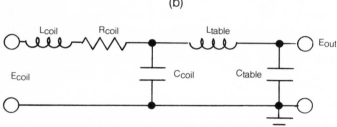

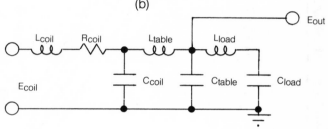

Fig. 14.12 (a) Changes in armature mechanical resonance with increasing load mass. (b) Electrical analog of (a); L_{table} represents stiffness between coil and table (damping not shown); C_{coil} represents coil mass; C_{table} represents table mass.

Fig. 14.13 (a) Changes in armature mechanical response due to SDoF load resonating around 500 Hz. (b) Electrical analog of (a); L_{load} represents K of sprung mass (damping not shown); C_{load} represents sprung mass.

14.4.2 Armature mechanical resonance. The high-Q resonance at f_4 in Fig. 14.9 is a mechanical resonance between (1) driver coil mass spring-connected to (2) the table and attached load mass. f_n drops when the load increases. At this high f_f, the armature no longer acts like a single mass; we represent driver coil mass by C_{coil} and table mass by C_{table}, which increases to represent increasing load.

At some frequency (around say 2,000 Hz) there will be a mechanical resonance similar to Fig. 12.5 (c) or (d). We represent this by a series electrical resonance between C_{coil}, L_{table} and C_{table}. Armatures with little damping (could be represented by a resistor paralleling L_{table}) have a very sharp ("Q" over 100) resonance. f_n drops as C_{table} becomes larger, representing increased mass load.

14.4.3 Effect of resonant loads. More realistically, consider Fig. 14.13. At say 500 Hz, C_{table}, L_{load} and C_{load} series resonate, representing a mechanical resonance between (1) table mass and (2) a sprung mass load. The sprung load responds vigorously (and may fail). See f_n, a sharp "notch" in table motion (input to the test load). The load absorbs energy from the table, thus reducing table motion; the load acts as a "dynamic vibration absorber." No doubt the operator will increase force — apply more electrical power to his shaker, endangering the load.

Much shaker force (observe large electrical power to the armature) is required to maintain constant motion. The load is prevented from (as it would in the "real world") affecting its own environment.

L_{load} and C_{load} act as series inductance and capacitance, shunting capacitance C_{table}. Below resonance, L_{load} and C_{load} behave as a net C. At resonance, $X_L = X_C$. Above resonance, they behave as a net L which resonates with C_{table}, resulting in f_p, a sharp "peak." (Throughout this discussion, we are interested in E_{out}, the voltage across C_{table}, representing table A. Readers may also be interested in the voltage across C_{load}, representing sprung mass motion.) Visualize the nearly motionless load assisting the table to vibrate, thus increasing table motion.

With a more complex load, each spring-mass resonance adds a notch and peak. Small mass-spring resonances will little affect a massive shaker table, but will greatly affect a small, light shaker table (more difficult to regulate table motion, if that is your goal). The peak and notch widths relate to the added load "Q". Often many such pairs, some large and some small, greatly complicate the graph of Fig. 14.13.

E_{out} for control *could* have been taken where L_{load} connects to C_{load}. That point represents a dynamic *response* point on the load, rather than its dynamic *input*. Some authorities prefer *response control* rather than the *input control* described throughout most of Sec. 14.4, 14.5 and 14.6. See Sec. I-4.2.6 of Method 514.3, MIL-STD-810D on "Input control vs. response-defined control". This compromises between, and possibly improves upon, the "infinite impedance" concept and the "zero impedance" concept (Sec. 14.8).

14.5 Manual intensity control too slow. Via his "force" control, the operator (Fig. 14.2) continually adjusts shaker A and D to meet older test specifications per Fig. 14.7. His indication of shaker motion is the vibration meter of Fig. 14.2. Meter delay + operator delay result in at least 1/2 second reaction time. With a Q of 50 at 500 Hz, f_n and f_p are only 10 Hz apart (at lower f_fs, even closer). On fast sweeps, only a few milliseconds may elapse between f_n and f_p; the human operator cannot react in time.

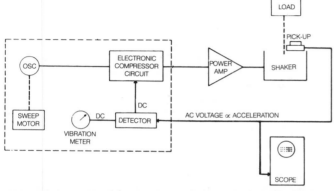

Fig. 14.14 Automatic analog control (in box) of shaker motion.

14.6 Electronic servo needed. Manual force control is too slow. We use an electronic compressor circuit (also called an AGC circuit) per Fig. 14.14. It regulates power amplifier input voltage to maintain A or D as desired. A "control" accelerometer senses table motion (monitor waveform by an oscilloscope). It serves to "close the loop" for control. The accelerometer signal also deflects the vibration meter. A casual observer might not be aware of resonances, as the meter reading remains constant. Show him the violent changes in driver coil voltage and current.

Assume we're maintaining 10g at the shaker. Our accelerometer sensitivity is 10 mv (peak) per g (peak). Thus 100 mv (peak) into the detector generates a request for 10g. If a resonance raises A to say 11g, as in Fig. 14.15(a), 110 mv into the detector will act to reduce motion to 10g. Should A drop below 10g, the reverse occurs. The system maintains any pre-set motion values quite closely.

Within the detector of Fig. 14.14 is a simple RC filter which minimizes test frequency ripple into the compressor. R and C values compromise two requirements:

1. Large time constant RC to smooth compressor voltage (even at low f_fs). Otherwise, the AGC would "follow" each cycle, distorting shaker motion.

2. RC must be small for quick response to vibration intensity changes; at high f_fs, mechanical resonances demand changes in a few milliseconds.

RC filtering should vary: sluggish at low f_fs and prompt at high f_fs. Figs. 14.3 and 14.16 exemplify modern analog oscillator/servo control units.

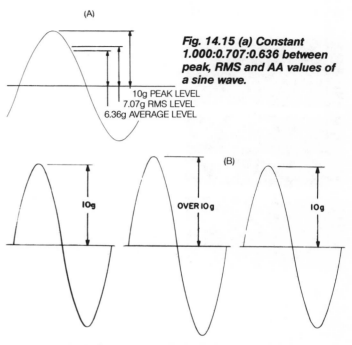

Fig. 14.15 (a) Constant 1.000:0.707:0.636 between peak, RMS and AA values of a sine wave.

10g PEAK LEVEL
7.07g RMS LEVEL
6.36g AVERAGE LEVEL

(b) Function of servo (human or electronic) is to maintain vibration intensity at for example 10g peak.

14.7 Need for tracking filters. With pure sine waves, there is a constant ratio between the peak, RMS and AA levels: 1.000:0.707:0.636. Thus the detector of Fig. 14.14 could sense the peak, the RMS, or the AA value. If one is held constant, all will be constant.

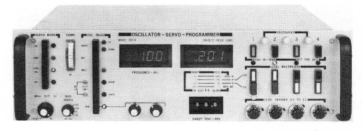

Fig. 14.16 Digitally generated sine source + analog intensity controller. Several intensity levels can be programmed. Frequency and intensity are displayed digitally. Courtesy Unholtz-Dickie.

However, various types of distortion may be seen on the monitoring oscilloscope, per Fig. 14.17. Distortion greatly upsets the 1.000:0.707:0.636 ratio. A servo can hold one level constant, but the others will vary widely. (Separate vibration meters, depending upon their detection circuitry, will also disagree, though they agree when calibrated upon pure sinusoidal voltages.) Further, an entirely different test results, depending upon which parameter is held constant.

Early San Diego sine vibration tests on Atlas missile electronic units consistently passed on a Calidyne system; they consistently failed on an MB system. The discrepancy was eventually traced to the servo detectors; Calidyne used a peak-sensing detector; MB an average-sensing detector.

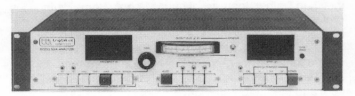

Fig. 14.19 Electronically-tuned filter can function as tracking filter for sine tests (also useful in random tests). Courtesy Trig-Tek.

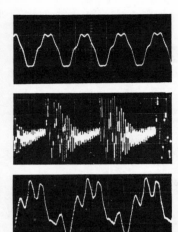

Fig. 14.17 Types of harmonic distortion seen at shakers during "sine" tests. Center example is due to parts "chattering," while others are due to power amplifier distortion.

Items should receive the "same" vibration test in any laboratory. Here is a device that might help: a tracking filter per Figs. 14.18 and 14.19 is electronically tuned by a reference signal from the oscillator. It insures that the specified test intensity *at the fundamental vibration test frequency* will be regulated. Unfortunately, harmonics and "hash" still affect the test item responses to some unknown degree, though they do not affect shaker force.

14.8 Infinite mechanical impedance. The real effect of servo control is to regulate shaker motion. No one objects to compensating for the shaker characteristics per Fig. 14.9. But should we prevent the load from affecting its own dynamic environment? Under servo control, the shaker table has "infinite mechanical impedance" (Sec. 6.3) as though it had infinite mass. Do we want this? Do we, at f_n (Fig. 14.13) where the sprung mass experiences greatest motion, wish to drive the shaker harder? This question might change test specifications and practices. (This question also applies to random vibration tests; do we really wish to equalize motion per Secs. 25 and 26?) Possibly *force* control would be better than *motion* control. Certainly the aircraft, or ship, or other structure to which our "black box" mounts in service *is* affected by the unit's response, does *not* have infinite mechanical impedance.

Whereas maintaining constant motion gives us an "infinite impedance" test, maintaining constant force would give us a "zero impedance" test. Results (on a "black box") would certainly be much different. See Sec. I-4.4 of MIL-STD-810D on "Mechanical impedance effects." Also Fackler in SVM-9.

Morrow has suggested using the former at low f_fs, where foundations tend to have higher impedance than do equipments, and using the latter at high f_fs, where equipments tend to have the higher impedances.

For the immediate future, "tailoring" tests per MIL-STD-810D should satisfy most requirements. If say a prototype aircraft and a prototype equipment are available, mount the equipment via force sensors and record their electrical signals for later analysis. For "input" control of subsequent tests, include both the peaks *and* the notches. Also measure the response motion at possibly critical response points within the equipment; you may decide to control motion at these points during tests, per Sec. 14.4.3.

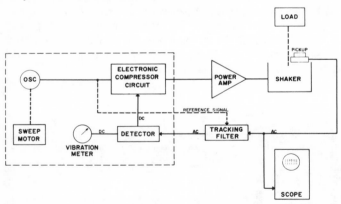

Fig. 14.18 A tracking filter has been added to the system of Fig. 14.14.

Section 15
Digital Computer Control
of Sinusoidal Vibration Testing

15.0 Introduction. Digital systems are mainly purchased for random tests per Sec. 26. However, digital systems can, for slight additional expense, also control sine and "sine + random" tests. Alternate "stand alone" digital sine controllers optimize dynamic range, control stability and compression rates for highly resonant structures.

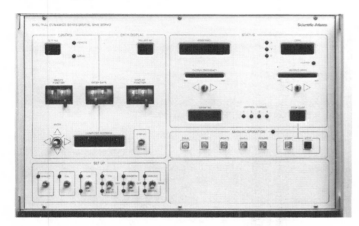

Fig. 15.1 Standalone digital sine controller with optimized control. Courtesy Spectral Dynamics.

Assuming you will use a system per Sec. 26, you need no additional gear for averaging, limiting, filtering, etc. (With analog controls these must be separately purchased and maintained.) Digital control also offers more levels of intensity, each with its own "abort" limit.

Some disadvantages: less convenient to set up or change D, A and f_c. For simple, brief tests, analog controls are easiest to use. (Be sure to read Sec. 14.4 re shaker and load dynamic responses.) However, digital sine controllers are gaining popularity due to greater reliability and accuracy. As with digital random control, the operator keyboards his test parameters, per Fig. 15.2.

Fig. 15.2 Typical dialogue between system and operator for setting up a digital sine test using computer control.

```
ENTER PARAMETERS? YES

INPUT  1=KYBD, 0=RT-11: 1

1 TEST ID: DET SINE
2 HEADING: DETECTOR SINE TEST
```

```
SWEEP PARAMETERS:
3 MODE 1=LOG, 0=LIN: 1
4 START, END FREQ, HZ: 10,5000
FREQ RANGE  OCTAVES, DECADES:  8.96 2.69
5 SPECIFICATION 1=RATE, 0=DURATION: 1
6 UNITS  1=OCT/MIN, 0=DEC/MIN: 1
7 RATE, OCT/MIN: 2
SWEEP DURATION  MIN, SEC:    4    28

TEST LENGTH:
8 SPECIFICATION 1=TIME, 0=SWEEP CYCLES: 0
9 CYCLES: 2
TEST TIME  HRS,MIN,SEC:   0    8    56

START UP AND SHUT DOWN:
10 START UP TIME, SEC: 10
11 SHUT DOWN TIME, SEC: 2

VIBRATION LIMITS:
12 DISPLACEMENT, IN(P P): 1
13 VELOCITY, IN/SEC: 70
14 ACCELERATION, G: 50

REFERENCE CONTROL SPECTRUM:
15 TYPE, VALUE, FREQ, ABORT LIMIT: 2,1,22,3
16 TYPE, VALUE, FREQ, ABORT LIMIT: 0,.02,40,3
17 TYPE, VALUE, FREQ, ABORT LIMIT: 2,3.2,100,3
18 TYPE, VALUE, FREQ, ABORT LIMIT: 1,2,300,3

19 TYPE, VALUE, FREQ, ABORT LIMIT: 2,10,3000,3
20 TYPE, VALUE, FREQ, ABORT LIMIT: 3,5,5000,3
21 TEST LEVEL (DB BELOW REF): 0

ACCELERATION CONTROL SIGNALS:
22 NR OF SIGNALS: 4
   CHANNEL NRS: 1,3,5,6
23 PROCESS 2=FUNDAMENTAL, 1=PEAK, 0=RMS: 2
24 SENSITIVITY, MV/G: 10
25 STRATEGY  3=AVG 2=RMS, 1=MAX, 0=MIN: 1

LIMIT SIGNALS:
26 NR OF SIGNALS: 2
   CHANNEL NRS: 2,4
27 PROCESS 2=FUNDAMENTAL, 1=PEAK, 0=RMS: 0
28 LIMIT VALUE,VOLTS: 1
29 LIMIT ABORT TOLERANCE (DB): 3

AUXILIARY SIGNALS:
30 NR OF CHANNELS: 1
   CHANNEL NRS: 1
31 PROCESS 2=FUNDAMENTAL, 1=PEAK, 0=RMS: 0
32 SENSITIVITY, MV/G: 10
33 MAXIMUM EXPECTED G: 20

TRANSFER FUNCTION:
34 REFEFENCE CHANNEL: 3
35 PESPONCE CHANNEL: 1
```

Continued on next page

```
MONITOR SPECTRUM:
36 MONITOR CHANNEL: 1

TRACKING FILTER:
37 TYPE 1=PROPORTIONAL BW, 0=FIXED BW: 1
   BW, %:70


ALARM LEVELS:
38 FREQ RANGE:  10,5000
39 +TOL (DB): 2.5
40 -TOL (DB): 2.5

41 COMPRESSION SPEED 2=HIGH, 1=NORMAL, 0=LOW: 1

REFERENCE LEVELS:
MAX DISPLACEMENT, IN(P  P):      .195
MAX VELOCITY, IN/SEC:      6.149
MAX ACCELERATION, G:    10
MIN ACCELERATION, G       .494
ACCELERATION RANGE (DB):    26.11


LIST? NO
CORRECTIONS? NO
SAVE? YES


DEVICE?: DX1
```

15.1 Setting up and running a test. The computer first asks the operator for a TEST I.D., (later needed for disk storage and future call up) and HEADING (for titling graphs). Then such parameters as log or linear sweep, sweep rate, duration, number of cycles, sweep limits, system limits, etc, are keyboarded (lines 3-14).

In Lines 15 to 21, TYPE requests codes for D, A and V. VALUE refers to those magnitudes. The FREQUENCY ranges and ABORT LIMITS are also keyboarded.

At Lines 22-25, the computer questions control accelerometer channels, their sensitivities and type of control (such as averaging, limiting, etc.). Here, four control accelerometers are to be monitored. MAXIMUM control is explained in Sec. 16.11. Lines 26 to 29 ask which accelerometer(s) could abort the test if their limit were exceeded. Thus any excessive response and/or input can stop the test. Lines 30-33 question the desired auxiliary channel(s) used for analysis or display during or after test. The transfer function between any two channels (operator selectable) can be displayed after a sine test, per Lines 34 and 35.

At Line 36, the operator chooses between control vs. response for "on line" monitoring. If, at Line 23, he had selected FUNDAMENTAL processing, then line 37 asks about a tracking filter: either proportional (percentage of center frequency) or fixed bandwidth.

At Line 38-40, he specifies alarm levels for various frequency ranges. At Line 41 he selects compression speed. (Some systems feature front panel controls.)

After the operator enters all values, the computer displays and prints data per Fig. 15.2. He should verify that his system can achieve all intensities. All data is saved (on paper tape or disks).

Now he presses his "start" button, sits back and watches. Intensity (at the starting frequency) slowly increases. If no feedback signal is sensed, the computer aborts and flashes the appropriate fault code or displays (depending on system used) "OPEN CONTROL LOOP" on the CRT.

If a feedback signal is sensed, intensity increases to full value and frequency sweep commences. During the test the computer continually monitors for any signal excesses, etc. Meanwhile, the operator can display (and make a hard copy of) such functions as control signal, response signal, drive spectrum and transmissibility. Some systems can present CO-QUAD and mechanical impedance displays, all fully annotated. Thus the operator and his "customer" can examine all aspects of the test during the test.

```
POST TEST DOCUMENTATION

DETECTOR SINE TEST

COMPLETION STATUS: NORMAL

TEST DURATION -- HRS, MIN, SEC:  0       8      55

MAX ABS CONTROL ERROR:     .91  DB AT 507.3 HZ.
AVG ABS CONTROL ERROR:     .082 DB.


CONTROL
CHANNEL          FREQ  RANGE (HZ)


SWEEP     1


   1              10.   --    18.72
   3              18.72 --    72.55
   1              72.55 --   148.3
   6             148.3  --   480.7
   5             480.7  --   812.1
   3             812.1  --  1235.
   5            1235.1  --  2240.
   6            2240.   --  2510.
   3            2510.   --  5000.


SWEEP     2


   3            5000.   --  2506.
   6            2506.   --  2233.
   5            2233.   --  1227.
   3            1227.   --   804.9
   5             804.9  --   478.7
   6             478.7  --   147.7
   1             147.7  --    71.78
   3              71.78 --    18.03
   1              18.03 --    10.
```

Fig. 15.3 Post-test documentation showing test heading, status, duration, error and control channel for certain frequency ranges.

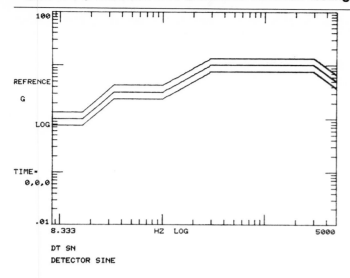

Fig. 15.4 Typical computer drawn reference line graph with tolerance lines.

Test results are hard copy printed and plotted for permanent documentation, as in Fig. 15.3 and 15.4.

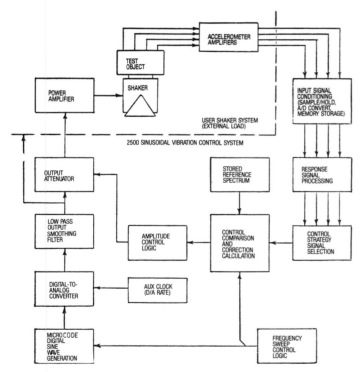

Fig. 15.5 Block diagram of a computer controlled digital sine system using four control accelerometers. Courtesy GenRad.

15.2 How the system works. See Fig. 15.5. Accurate, low distortion sine signals are computer generated over a frequency range of 0.1 to 5,000 Hz. Table motion or specimen responses can be controlled. Algorithms select the peak

value, compute the average or the RMS value or digitally filter signals; actions resemble tracking filters per Sec. 14.7. We can control either the largest, smallest or average of the control channels.

The measurement process output is actually a table of numbers. These represent the control channel error (or the error of each of several channels). Under AVERAGE control, the system averages the control channel errors. Under MAXIMUM control, the largest numerical error is selected and amplitude controlled.

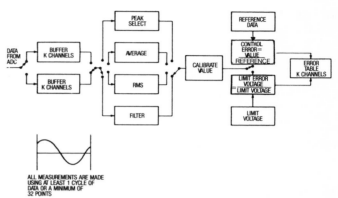

15.6 Block diagram showing four measurement techniques for digital sine control. Courtesy Hewlett-Packard.

15.2.1 Four measurement techniques are available, per Fig. 15.6:

PEAK SELECT: Data is searched for the absolute peak value, which is divided by the operator-set reference level. This ratio goes into the error table. Use this mode when waveforms are fairly clean. If waveforms are distorted, the peak value varies, causing control to appear erratic.

AVERAGE: Rectified values are summed, then divided by the number of samples. This average value is multiplied by 1.57 (peak:average ratio for sine), then divided by the reference for the error table. Results are stable even when distortion creates errors.

RMS: Sampled values are squared and averaged, then square rooted, then multiplied by 1.414 (peak:RMS ratio for sine), then divided by the reference for the error table. This control method emphasizes major motions, and is useful when preventing overtest.

FILTERING: This single frequency FFT ignores harmonic distortion and hash; it senses only the fundamental. This value is divided by the reference for the error table.

15.2.2 Control strategy. Following measurement, the error selection software uses the Line 25 entry. Individual limit channel(s) are monitored.

If the controlling strategy is AVERAGE, the computer averages the errors in the channels. This average and the channel errors are scanned. Per Fig. 15.7 the largest numerical value is exponentially averaged, to be compared to the alarm and abort values. If "alarm" is exceeded, the computer notifies the operator; if "abort" is exceeded, the system smoothly shuts down. If the average is OK, all signals are available for the amplitude control algorithm and the output attenuator.

If the controlling strategy is MAXIMUM, all errors in the table are scanned and the largest absolute error is selected. If the controlling strategy is MINIMUM, the smallest accelerometer signal is compared to the limit channel(s). The largest limit and control channels are selected and if either exceeds "abort," the system shuts down. Whichever control strategy the operator selects, the system protects against "runaway" conditions.

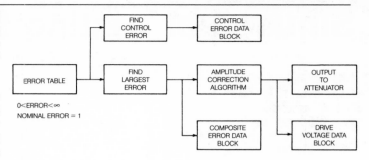

15.7 Block Diagram showing error table and control. Courtesy Hewlett-Packard.

Section 16
Sinusoidal Testing Standards

16.0 Introduction - purpose. Here in Sec. 16 we consider sinusoidal vibration tests, seldom specified today. Random tests (Sec. 23), intense noise tests (Sec. 31) and shock tests (Sec. 33) are today preferred. The following statement of purpose is taken from MIL-STD-810D dated 19 July 1983. "Vibration testing is performed to determine the resistance of equipment to vibrational stresses in its shipment and application environments." Sine tests are still permitted under Sec. I-4.2.3 of -810D. The following -810D paragraph will interest readers:

"1.1 *Purpose.* This standard provides:

A. Guidelines for conducting environmental engineering tasks to tailor environmental tests to end-item equipment applications.

B. Test methods for determining the effects of natural and induced environments on equipment used in military applications."

The following statement tells when this standard is to be used.

"1.2 *Application.* Application of this standard early in the development phase of the acquisition process is encouraged. Selected application at other points in the acquisition process may be appropriate. The methods of this standard are not all-inclusive. Additional environments or combinations of environments should be included in the environmental test specification when appropriate. The test methods of this standard are intended to be applied in support of the following objectives:

A. To disclose deficiencies and defects and verify corrective actions.

B. To assess equipment suitability for its intended operational environment.

C. To verify contractual compliance."

This and comparable standards are issued by various agencies and firms in the USA and abroad. Standards appear in requests for proposal (RFP) to warn suppliers that their firm must prove their products are acceptable. Standards describe tests that must be performed.

-810D is a refreshing change from -810C (and most other standards). Most require tests to specific environmental extremes, without explanation as to data sources, and without opportunity to effect changes when needed. -810D requires "tailoring" to individual applications.

Such standards can become part of a contract. Rather often the "customer's" inspector witnesses tests. The manufacturer may contract testing to commercial laboratories. (Sometimes the military service involved will perform some of the tests.) If an item fails a test, it must be improved and tested again and again until it passes.

Someone within the firm planning to supply an equipment must understand the standard or specification to be sure that the firm or an outside lab can perform the required tests. Product design should be followed by some developmental testing, so the product will surely pass all contractual tests.

The following is taken from Method 514.3, Vibration, of MIL-STD-810D. "PURPOSE. Vibration testing is performed to determine the resistance of equipment to vibrational stresses expected in its shipment and application environments."

16.1 Tests do not simulate the real world. How did such vibration standards come about? It appears simple to measure real world vibrations and then simulate these. But difficulties arise, especially if only a few samples will be built and tested. Possibly little or no in-service data is yet available; somewhat hypothetical preliminary tests must be devised in order to specify equipment. But even where in-service measurements *are* made, resulting data can be severely mishandled. (See also Sec. 6.7 and 14.8.)

16.2 An example. One classic example of in-service testing to develop vibration testing specifications: vibration was surveyed on various classes of military aircraft, during various maneuvers, carrying various loads, under various flight conditions. At different structural locations, sensors generated signals which were on-board oscillograph recorded. The records were later visually analyzed; displacements (D) at various frequencies were determined and plotted as in Fig. 16.1. Finally, an envelope was drawn around most of the data points. This envelope, defining D from 10 to 50 Hz, and defining A from 50 to 500 Hz, became part of a standard; its authors selected a sweep speed and total test time.Note that enveloping peaks ignores notches and there causes extreme overtest (see Sec. 14.8). Possibly a better test would result from (in flight) measuring *force* between aircraft and actual loads. But until recently, force sensors per Sec. 6 were not readily available, whereas motion sensors have long been available.

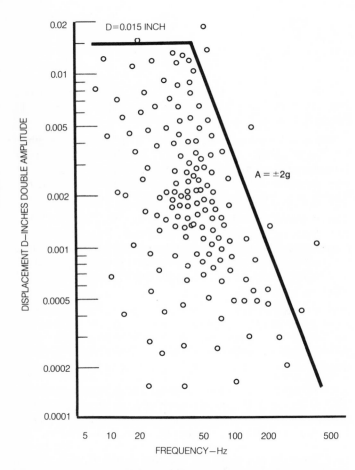

Fig. 16.1 Graph of early sinusoidal vibration environmental test, with 0.015 inch D from 10-50 Hz and 2 g peak A from 50-500 Hz.

Very likely the authors would have preferred constant A down to 10 Hz, rather than change to constant D, had sufficient shaker D been available. (See Sec. 14.2 re crossover frequency.)

Such tests are not close to flight conditions. Note in Fig. 16.1 that the envelope is more than 100 times as severe as some flight vibration that was observed. The curve encloses 95% of the measured data points (it excludes the highest, questionable 5%). Hopefully such an arbitrary committee-developed test uncovers poor design and/or poor assembly.

By cycling 10-500-10 Hz, all resonances are sequentially excited (though sweeping too fast will not fully excite each resonance). Fatigue damage will occur if resonance is severe enough and/or lasts long enough. Specimens having many severe resonances will suffer more damage than will those having few, highly-damped, resonances. Such a simple test can be administered with some uniformity in many test labs.

16.3 Tailoring tests is encouraged by MIL-STD-810D. "Tailoring" requires that the test engineer know the test items's future usage or transportation environment. Where does he get that information? It should be available from the involved military service and/or the prime contractor (long before testing commences) at the time of the RFP or RFQ. On prime contracts, the services are obligated to provide data on which to base tests, based on field measurements during actual usage or shipping. What if the vehicle (on which the test item will be attached) has not yet been built? Someone (military service or prime contractor) must predict the vibration. Tailoring is actually an old idea; nearly all MIL standards tell you "If you have environmental information, use it." But until -810D, most standards showed "canned" test programs, and many laboratories used these, even when clearly not appropriate.

16.4 Classes of equipment. Specifications sometimes classify tests for different classes of equipment, such as:

1. Aircraft
2. Helicopters
3. Air Launched Missiles
4. Ground Launched Missiles
5. Ground Vehicles
6. Shipment by Common Carrier, Land, Sea or Air
7. Ground Equipment (excluding ground vehicles)

Sinusoidal vibration of flight equipment was desired for many years, with constant D at the lowest f_fs, then constant A at higher f_fs, 500 Hz or 2,000 Hz, depending upon aircraft type, as in Fig. 14.7. Various sine tests for equipment installed in ground vehicles call for f_fs from 2 to 500 Hz, with D to 9 inches and A to ±2.5 g. Still other tests are prescribed for equipment to be transported by railroad, etc. Some U.S. Navy shipboard equipment tests swept to 33 Hz, others to 50 Hz. Random vibration has long been specified for missiles and high-performance aircraft components, and is now specified for nearly all applications.

16.5 Will shaker perform test? When you first see a standard (either before a proposal or soon after contract award), carefully read it and any detailed specifications that apply. Clarify any hazy portions long before testing. Determine that your shaker system is adequate. Have you sufficient D without external support (vertical vibration)? Is horizontal vibration required? Does the system produce enough force per Sec. 12.3? If the crossover frequency f_c is not stated, calculate it. Check V at f_c, per Sec. 12.6.

16.6 Fixture. Design, build and experimentally evaluate a suitable test fixture to hold the test item approximately as it is held in actual service, per Sec. 19.

16.7 Purity of motion. The goal of standards such as MIL-STD-810: that the test item receive (along one axis at a time) sinusoidal or random input motion (pure translation—no rocking) at specific frequencies or over specific frequency ranges and at specific A or D or spectral density levels. Unfortunately, numerous variables interfere with achieving that goal. Let us eliminate all possible variables, so our test results will correlate with results from other laboratories.

Use an oscilloscope to observe control accelerometer waveform (unfiltered signal). Remedy the cause of any non-sine motion (structural failure or loosening of parts in the test item, the fixture or the shaker itself; faulty accelerometer and/or cable; harmonic-excited resonance due to power amplifier distortion). A tracking filter (Sec. 14.7) helps to eliminate testing variables.

16.8 Resonance search. Finally, mount your test item on the fixture (already attached to the shaker). Many standards require searching out test item resonances. Detect them by ear, by fingertip, by accelerometer waveform on your 'scope, by increased vibration intensity as noted on a vibration meter, by change in power demand from the power amplifier, by changes in test item performance, etc. Investigative logic will help to decide which f_ns are "critical" and must be reported. (Some standards require long-time testing at each critical f_n, so don't report these unless they actually exist.)

Must the test specimen function during vibration tests? Move specimen support equipment close to the shaker and connect it to the test item. "Project" people should operate this auxiliary gear and watch for test item malfunction so you can concentrate on test control. Should the test item malfunction, perhaps you'll stop the test for remedial action.

The foregoing is very abbreviated, of course. We cannot give sufficient detail for all of the varied circumstances you will encounter during vibration testing. Some of the problems (and solutions) commonly found are described below.

16.9 Stroboscopic light. Resonance search is aided by a "strobe" light whose flash rate is equal to (f_f - 1 Hz). The specimen appears to move at 1 Hz. Parts responding with greater amplitude than their neighbors immediately catch your eye. Such visual observations supplement measurements because

  1. Sensors cannot be placed on all components (resistors, capacitors, relays, transistors, PCBs, etc.) due to sensor mass and difficulty of attachment.

  2. Direction of maximum vibration and node/antinode locations are seldom known in advance.

We might miss instrumenting some important element. The strobe light helps to identify resonances. If an element responds at *other than 1 Hz*, investigate further to find the cause, such as power amplifier distortion.

An assistant could manually and continuously adjust the strobe firing rate to (f_f - 1 Hz). This is difficult since he cannot see small Ds off-resonance. Better to synchronize flashing per Fig. 16.2 and 16.3. Specialized movie and video cameras record responses for design improvement.

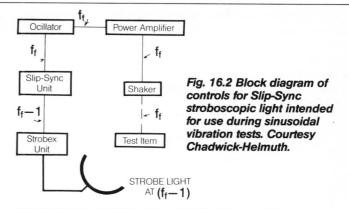

Fig. 16.2 Block diagram of controls for Slip-Sync stroboscopic light intended for use during sinusoidal vibration tests. Courtesy Chadwick-Helmuth.

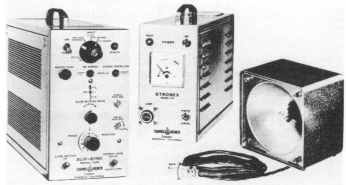

Fig. 16.3 Hardware to implement Figure 16.2. Courtesy Chadwick-Helmuth.

16.10 Single control accelerometer difficulties. With one "control" accelerometer, we can control motion at one specimen attachment point. But additional accelerometers will show us we are not controlling motion at the others. Imagine that at each f_f (as in Fig. 16.4) we plot the greatest and the least of our "input" accelerations. Note the tremendous differences at certain f_fs due to specimen and/or fixture resonances. (Regulating those differences is the purpose of column 4, Table 19.1.) Visualize how different would have been that test had another "input" accelerometer controlled, rather than the one which *was* used for control, which *was* held at 2 g. At 350 Hz shaker intensity might have increased 10 times; at 85 Hz it might have halved.

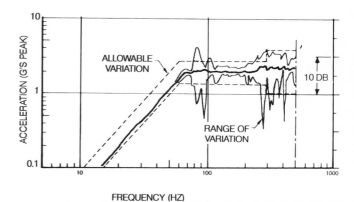

Fig. 16.4 Electronically averaged signal was maintained at 2 g. Others varied 0.2 to 4.5 g. Testing under control of a single accelerometer would yield vastly different results.

16.11 Multiple accelerometers, selecting or averaging

offer solutions. Consider four accelerometers to monitor motion at, say, four specimen input points. They may differ by say 10:1. Who decides which accelerometer "controls," sending a signal to the servo unit of Sec. 14.6? And on what basis does he decide? The decision greatly affects test outcome.

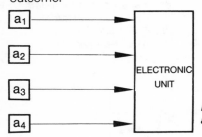

Fig. 16.5 Use of selector or averager in control loop.

Instead of a human decision, pass the four signals to an electronic unit, per Fig. 16.5. It could select the largest for the servo system. Thus, on a 2 g test, one attachment point experiences 2 g and the others experience less. The likelihood of failure is small. This device helps "pass" tests by a form of cheating. Another possible application: control vibration intensity with a_1 monitoring acceleration in the desired direction and a_2 and a_3 monitoring lateral or rocking motion. If either a_2 or a_3 exceeds 2 g, it will "control." Whatever the reason (shaker or test load resonating) for unwanted motion exceeding desired vibration, such control is not intended by the specification writer. The authors disapprove of this device *and* of a similar strategy discussed in Sec. 15.2.2.

More logical: let Fig. 16.5 represent an averaging circuit. Then the servo control signal represents the average of four ac signals

$$\frac{a_1 + a_2 + a_3 + a_4}{4} \qquad (16.1)$$

Beware of phase differences such that the average ac signal is zero. The servo unit will "think" the shaker is not vibrating and will immediately increase electrical power to the shaker. One solution: rectify the four ac signals into four dc signals and average these to ignore phase. Or use time-sequence averaging per Fig. 16.6. The unit samples one period (T = 1/test frequency) from each control signal, then servo-controls the A value of the composite (sampled) signal. Some attachments will experience more (and some less) than 2 g. When this averager is used for random vibration, it

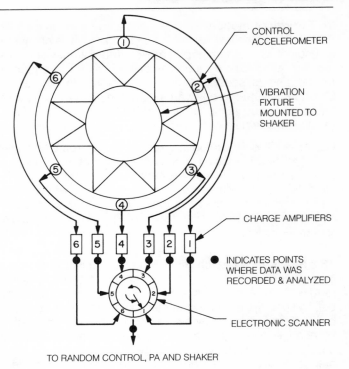

Fig. 16.6 Time division multiplexing (MUX) among several accelerometers. Courtesy NASA.

samples equal intervals from each sensor. The ASD of the composite (sampled) signal is then controlled manually or automatically. In each frequency window, some accelerometer ASD's will be greater (and some less) than the controller-set ASD.

Selectors and averagers can also be used in protective modes. Multi-accelerometer averaging can prevent excessive testing due to specimen and/or fixture resonances, or excessive lateral motion. Note that loss of one accelerometer signal won't cause "runaway" as with single point control. Digital equivalents of these devices are discussed in Sec. 27.5.

16.12 Standardization between labs is minimal. Wide test method variations exist between laboratories. Various methods have merit, but they accomplish different results. A test item may "pass" at one facility, only to "fail" at another. "Standards" are not very explicit.

Section 17
Conducting Sinusoidal Vibration Tests

17.0 Introduction. Though the amount of sinusoidal testing has dropped rapidly (in favor of random testing), some is still required. We have examined (Sec. 14 and 15) the equipment to be used, and looked at test standards (Sec. 16). Now let us consider how to conduct various types of sine tests.

17.1 Fatigue tests. Shakers can be used for fatigue tests, to determine certain material properties. For 100 years, fatigue testing utilized single-frequency "brute force" mechanical shakers per Sec. 10. However, electrohydraulic (EH) and electromagnetic (EM) shakers increasingly are used for random fatigue tests.

To perform a fatigue test, we will attach many identical samples, one at a time, to our shaker. The mounting depends upon the desired loading: tension, compression, or both; torsion; or bending. We will vary intensity of applied vibratory force to control peak stress level in the samples.

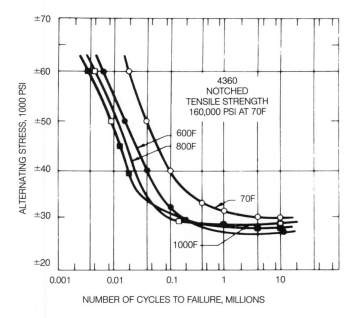

Fig. 17.1 "S−N" curve from fatigue testing. Severe loadings (short life) might be suitable for missiles. Light loadings (long life) might be suitable for passenger or freight aircraft.

The harder we stress a sample, the fewer cycles it will endure before breaking. Fig. 17.1 graphs peak stress S per cycle vs. number of cycles N to failure: an "S-N" curve. Designers use such curves to decide how much stress is permissible for a needed structural life. Recognize, however, that service and transportation vibrations are never sinusoidal; thus these tests do not indicate performance in the "real world." Note also that our carefully machined samples do not represent mass produced parts.

Samples are often "notched" (deliberately weakened at a defined point) so stresses will be high; here we will place strain gages. On complex samples some preliminary testing would be required to locate an existing high stress point.

Assume a test frequency of 100 Hz. We will regulate force to maintain the desired S until failure occurs. Then we will calculate $N = f_f \times$ time. At each S, and at various ambient temperatures, we will test many samples to failure.

If our EM shaker does not develop sufficient force, a resonant coupling can apply greater force. Rather than an arbitrary frequency such as 100 Hz, we can tune the system so resonance aids us. As time elapses (during a fatigue test), S gradually reduces; more shaker force is needed as failure approaches. Slight cracks and localized failures may be invisible, but (1) they add damping so that "Q" is reduced; also (2) f_n changes. We may need to "track" the shifting f_n right up to the instant of failure.

Such testing should be automated. Several firms offer resonant dwell systems. As failure approaches and f_n shifts, f_f should also shift. Additional environments may be required.

17.2 Transmissibility investigations of, for example, vibration isolators (for protecting delicate equipment against damaging vibration and shock) may require a Tr vs. f_f curve similar to Fig. 1.4. At very low f_f, input and output motions are equal, and Tr = 1. At resonance, Tr reaches a peak value called "Q". Above resonance, Tr drops off.

Q and f_n are both affected by the input force or drive level. If we test at 5g, for example, the maximum response may be 50g (Q = 10) at say 42 Hz. But if we test at only 1g, Q may be

12 at 45 Hz. We should perform all Tr tests at specified intensity, so that data, perhaps from different test laboratories, may be fairly compared.

Another example of Tr testing: evaluating vibration test fixtures prior to use. Bolt the empty fixture to the shaker. Attach accelerometers to measure responses at different points and in different directions. With an ideal fixture, all axial motion equals the input (Tr = 1.0) up to perhaps twice the top f_f. Transverse accelerometers should read zero. Tr plots aid the fixture designer on future designs.

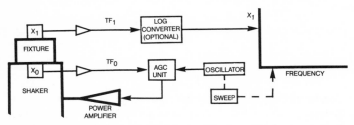

Fig. 17.2 Automatic analog transmissibility Tr plotting to evaluate fixture at servo-controlled 1g intensity. TF_0 and TF_1 represent tracking filters in the control and response signal paths (see Fig. 14.18).

We'll save time (in evaluating test fixtures and other dynamic structures) by automatically plotting Tr, rather than manually taking data, calculating and plotting. One simple method is shown in Fig. 17.2. An AGC circuit (see Sec. 14.6) regulates input motion at 1g (or other constant value). The response signal is converted to dc for X-Y plotter Y input. With the input constant at 1g, an equivalent to Tr is plotted. The ac-dc converter is often a "logarithmic converter" so Tr is plotted on a "decibel" or logarithmic scale.

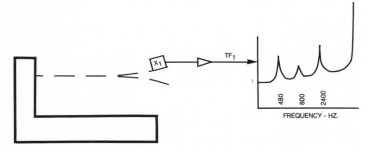

Fig. 17.3 Shaker harmonic distortion can excite the valid 2,400 Hz fundamental cantilever response at odd subharmonics, yielding invalid resonance data. A tracking filter at TF_1 will prevent invalid peaks from appearing on the Tr curve (which should only have one peak, at 2,400 Hz.)

Plots per Fig. 17.2 and 17.3 are facilitated by tracking filters TF_0 (see Sec. 14.7) in the control loop, and TF_1 for response measurement (optional but very helpful). TF_0 and TF_1 insure that measured and plotted quantities represent Tr *at the oscillator frequency*, unaffected by harmonic distortion, "hash" and other sources of error. They prevent meaningless extra peaks on our graphs (see Fig. 17.3) and save having to explain those peaks to our clients.

17.2.1 Resonance searches upon a structure determine the particular f_fs at which resonance occurs: at which Tr peaks. Recall the simple beam of Fig. 1.10; resonance occurred at several frequencies.

Imagine a resonance search upon a horizontal test plate held and driven by its edges. We sprinkle cigarette ashes, fine sand, or salt on the plate; at resonances the powder will collect along the lines of least vibration: the nodal lines. This is elementary modal analysis.

Fig. 17.4 A relatively simple structure. Yet a resonance search will identify dozens of f_ns.

Structures are usually composed of various plates and beams, each with its own vibration characteristics. Consider the structure in Fig. 17.4. If we test it 10 to 2,000 Hz, and carefully watch every element, we will find dozens of resonances. Masses and stiffeners (to help carry concentrated loads) create dozens more resonances.

17.2.2 Critical frequencies. Some resonances are dangerous because continued vibration may lead to failure. These are the f_ns we seek via resonance searching. Our customer should advise us what kinds of resonances or critical frequencies are of interest, and, in general, how to observe them. For example, a design engineer may need the critical f_ns of the deck in Fig. 17.4, for defining the environment of components. (He won't care about outer surface resonances.) So we'll mount accelerometers on the deck. During a frequency sweep, we'll record many f_ns. A test plan which merely says "record all resonant frequencies," but does not tell which are important is vague and ambiguous.

We must be sure (when reporting critical frequencies) that responses occur at the shaker driving frequency f_f. Shaker distortion may cause responses at multiples of shaker frequency. For our test the oscillator might be set at 100 Hz, but motion might be distorted and contain not only 100 Hz but also harmonics at 200, 300, 400, 500, etc. Hz. If this *complex* vibration causes a deck resonance, we must identify which is critical. Visualize an oscilloscope showing deck acceleration; the signal contains 100 Hz and 300 Hz, similar to the upper right quadrant of Fig. 13.6. We temporarily record the acceleration Tr we see on a meter. Later, when our f_f is 300 Hz, our 'scope shows pure sinusoidal motion, with a higher "Q" than before. We permanently record 300 Hz as the f_n. Recall our need for low PA distortion, per Sec.13.6.

Simultaneous motion at sub-multiples of f_f (such as 50 Hz or 33-1/3 Hz when testing at 100 Hz) is usually caused by test load non-linearity. This complex subject is beyond the scope of this text, but note that it can cause surprising test results. Resonance can occur at different frequencies when scanning upward than when scanning downward. Resonances may or may not appear, depending upon vibration intensity.

Always use an oscilloscope to monitor the accelerometer signals from important structures. If vibration contains components (distortion) at a frequency different from the input, note that fact and investigate why. Never trust meter readings with distortion present, unless the signal is "cleaned up" ahead of the meter by a tracking filter. Without these precautions, you may unwittingly report critical frequencies that do not exist.

17.3 Qualification tests (upon pilot-production models of a new unit) prove capability of exceeding expected environmental requirements. Thus such tests are run at higher test levels for longer periods than on later production tests. Such tests are very important, because failure or marginal behavior may lose a contract.

17.4 Production acceptance tests are run upon some percentage (up to 100%) of the items (component or assembly) being built. Tests are usually less severe than in "qual" testing, and generally shorter.

17.5 Transportation tests can be performed with sinusoidal vibration upon mechanical shakers; see Section 10.4. Better to utilize EH shakers per Sec. 11.5 and more realistic random vibration per Sec. 21 through 27.

Section 18
Orthogonal Motion

18.0 Introduction. Orthogonal motion (also called transverse, cross-axis, quadrature and lateral motion, also cross-talk) usually refers to some departure from idealized single-axis motion. Here we discuss its causes, its measurement and how to reduce it for single-axis vibration testing. See also Sec. 5.16.7 re lateral accelerometer sensitivity.

18.1 Single-axis tests. Most specifications call for single-axis (at a time) motion. Some specifications call for extra-severe vibration in the environment's most severe direction. We should limit the amount of cross-talk (to 10%, 30% or even 100% of the desired motion) to control and standardize tests. The vector sum of ±10g vertical plus ±10g horizontal vibration is ±14.1g. Added motion at a critical horizontal f_n could cause undue damage in laboratory A but not be present in laboratory B.

Fig. 18.1 Describing solid-body motion requires three orthogonal and three rotational axes.

18.2 Single-axis motion. Imagine a shaker armature and attached load, moving per Fig. 12.5(a). The load is homogenous and has no resonances. The load cg is aligned with the shaker armature. All flexures have equal stiffness. All motion is along the vertical axis X, Fig. 18.1. There is no unwanted transverse motion (no Y or Z) and no rotation in any direction (no Φ, Θ, or Ω). Of the six possible degrees of freedom (three linear and three angular), all motion is zero except X. Such idealized motion is rare; we must measure what really occurs.

18.3 Shakers not single-axis. Even with no load, some unequal flexure stiffnesses and/or imperfect armature balance can cause rocking per Fig. 12.5(b). Whereas shakers of

the 1950's had over 100% cross-talk at certain frequencies, today one can specify cross talk under 10%, no load.

See Sec. 12.7; individual flexures having different dynamic stiffness K contribute to unwanted rocking. One user removed the flexures from four identical shakers, then matched and reassembled, reducing cross-talk 3:1. The armature must be dynamically "balanced;" its cg must lie along the table axis. Electrical and/or coolant connections must be symmetrical.

18.4 Lateral motion checks should be included in acceptance testing of a new or reworked shaker. Mount a triaxial accelerometer assembly at table center. With sinusoidal X motion, say ±10g, slowly sweep through the frequency range. If Y or Z motion is excessive, reject the shaker. Remember to allow for cross-axis sensitivity in your Y and Z accelerometers. (See Sec. 5.16.7.) Verify that any accelerometer mounting block is rigid at all frequencies and is firmly attached, or you may get erroneous readings. Per Sec. 5.16.3, "homemade" mounting blocks, particularly of plastic, can give trouble.

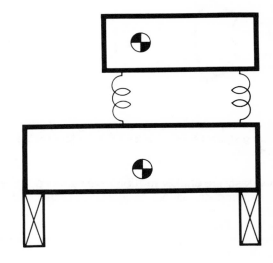

Fig. 18.2 Statically-balanced but dynamically-unbalanced EM shaker armature. Straight-line motion is unlikely.

18.5 Eccentric loads. Fig. 18.2 suggests vibration isolators under the load. It can easily move in all directions relative to the shaker table. (Of course, small motions can occur even with "rigid" bolted or welded connections.) The load is static balanced (cg is along the armature axis, but perhaps not midway between the isolators). At certain f_rs, the load will rock as well as translate in the X direction. Load rocking and rotation (especially if large and heavy) is an unwanted "mechanical feedback" to the armature. Centering your load cg is no guarantee of straight-line dynamic motion.

Now consider a "black box" of components, chassis, structural elements and cover plates, connected via electrical cables and/or hoses (probably unsymmetrical) having both mass and stiffness. Various parts of the load now resonate in various directions. Some show more lateral motion (Y and Z directions) than in the desired X direction.

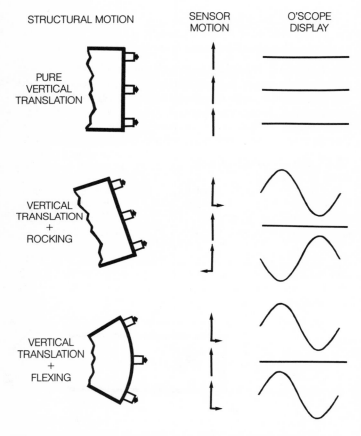

STRUCTURAL MOTION   SENSOR   O'SCOPE
                           MOTION   DISPLAY

PURE
VERTICAL
TRANSLATION

VERTICAL
TRANSLATION
+
ROCKING

VERTICAL
TRANSLATION
+
FLEXING

Fig. 18.3 Pure vertical translation vs. rocking vs. flexing and resulting signals from idealized horizontal-sensing accelerometers.

18.6 How describe shaker motion? We cannot completely describe any motion without concurrently measuring both amplitude and phase at two or more points. Consider the body shown in Fig. 18.3, with nominal X motion. Three idealized accelerometers having zero lateral sensitivity (see Sec. 5.16.7) sense any horizontal vibration; all have zero output. If the body rocks or bends the accelerometers experience different relative motions. You need a phase meter or (per Sec. 18.9) Lissajous' pattern techniques on an oscilloscope.

Phase comparison identifies translation vs. rocking (whole-body motion) vs. bending.

Use these techniques when evaluating shaker and/or fixture motion. Apparent cross-talk may not be horizontal translation or rocking but rather may be flexing. This is particularly true with "table diaphragming" or "oil-canning" per Fig. 12.5(c). The driver coil connects more rigidly to the rim than to the table center. Motion is solid body at low f_rs. At high f_rs the center has perhaps 50 times greater x and $\ddot{x}$ than does the rim. (With the armature of Fig. 12.10, this mode occurrs around 2,000 Hz.) At higher f_rs (usually above the advertised range) the table acts as a vibration isolator (see Sec. 1.1.5); the center motion is less than at the rim. Ideally, the entire table area has pure X translation so that all mountings of a large test item (or several small items) will receive the same motion. We need several accelerometers. If we instrumented only one point (Fig. 18.3) we might not recognize rocking or flexing.

18.7 Difficulties with small shaker. Small, light resonant loads do not cause much cross-talk in a relatively large heavy shaker armature, but can rock the armature of a small shaker. Testing difficulties and delays may cost more than buying a big shaker. Avoid test loads (fixture + item) exceeding half the armature mass; excessive cross-talk, even broken flexures, may result.

Sometimes even the largest available shaker is too small, as for the rocket motor of Fig. 18.4. Test load mass greatly exceeds armature mass. Aside from the dead-weight problems of test load support (see Sec. 20.2), load rocking endangers the shaker.

More typically, consider a 34 kg (75 lb) armature that develops 22,250 N (5,000 lb) vector force. With a load of 17 kg (37 pounds) this shaker will develop around 440 m/s² (45g) peak. You may be tempted to vibrate loads around 190 kg (425 pounds) at say 10g or even heavier loads at 5g. Such large loads may even "overhang" the table. Rocking may result, and may loosen or damage the suspension attachments; this risks future test quality, even with small loads.

18.8 Shaker restrains load motion. Consider a resonating load, driven vertically. Some resonant element creates horizontal force. If horizontal motion is restrained (by a "stiff" armature and flexures) the load will respond much differently than if tested on a "soft" armature and flexures (or on a piece of aircraft structure per page 514.3-33 of MIL-STD-810D.)

18.9 Lissajous patterns viewed on cathode-ray oscilloscopes can measure (1) phase angles between two (same frequency) sine waves and (2) frequency ratios between sine waves. The two signals can represent two motions or two forces.

18.9.1 Determining phase angle. Per Fig. 18.5 apply one signal to the "vertical" or "Y" input terminals to move the cathode ray in that direction. Apply the other to the "horizontal" or "X" input terminals to move the cathode ray in that direction. (Disable the built-in horizontal sweep.) To find the relative phase angle θ between the two signals, adjust the Y and X amplifications so they are equal (same CRT dimension on a "trial" signal). Points from 1 to 6 represent various times

Fig. 18.4 Some test items dwarf even the largest EM shakers. Courtesy Wyle Labs.

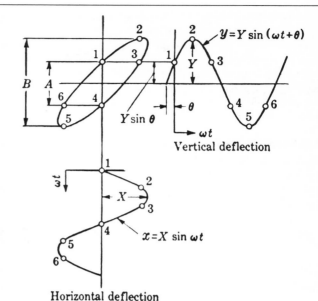

Fig. 18.5 Two same-frequency sine signals combine to form an elliptical Lissajous pattern on an oscilloscope. Here the signals have the same frequency but somewhat different phase.

within one cycle. The X-axis signal is our reference; the Y-axis signal leads by an angle θ. At the moment of zero X, Y has already progressed to angle θ, and has reached a height of $Y \sin\theta$ (point 1). Thus the intercept of the ellipse with the vertical axis is $Y \sin\theta$; we measure $Y \sin\theta$ on the 'scope face, and then look up angle θ in trigonometric tables. Use of the double intercepts A and B reduces errors; sine θ equals A/B. Unfortunately, (1) we have trouble knowing whether θ represents a phase lead or lag; (2) accuracy decreases rapidly above 45°.

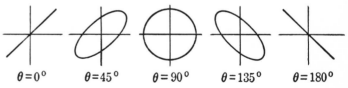

Fig. 18.6 Lissajous patterns for several phase angles.

If a vibrating structure (such as a shaker table) moves in a straight line (zero rocking), the pickups will move in-phase and the result will resemble the 0° sketch in Fig. 18.6. However if that 45° line changes shape, phase shift, possibly exceeding your shaker acceptance criteria, is indicated.

Seeking a structure's nodal pattern, while it is being driven by a shaker at some f_n? You need the locations of all nodes and antinodes, but perhaps have relatively few sensors attached. Examine θ between adjacent pairs of pickups. (If a true resonant mode, all θs will be 0° or 180°.) If the signals from adjacent sensors are 180° apart, a node lies between them; perhaps you can estimate its distance from each pickup.

Suppose you need to accurately determine an f_n. We'll define resonance as a 90° phase shift between (1) the signal from a force sensor (see Sec. 6) located between shaker and specimen, or from the shaker driver coil current, and (2) the signal from an accelerometer located at some structural point of interest. At resonance, your 'scope figure will resemble the center pattern of Fig. 18.6. You can measure phase angles more accurately with a "phase angle meter," if available. But any lab having a shaker has at least one 'scope.

18.9.2 Identifying frequency ratios. Do two points on your structure vibrate at exactly the same frequency? If your scope pattern (fixed) resembles Fig. 18.6, their frequencies are equal. A slow rotation indicates a slight frequency difference. A large frequency difference is shown by a rapid roll.

Fig. 18.7 Lissajous patterns for several discrete frequency ratios between vertical and horizontal signals.

At certain discrete frequency ratios, the 'scope pattern resembles Fig. 18.7. To read frequency ratios between the horizontal and vertical signals, count the "points of contact" between the pattern and imaginary horizontal and vertical boundaries. When the pattern makes two vertical traces for each horizontal trace, the vertical frequency is twice the horizontal. Sometimes points on a structure move at different frequencies, such as 2:1 or 3:1. This often results from a distorted (non-sinusoidal) input force driving a structure having several f_ns.

Consider evaluating horizontal motion (ideally zero) of a vertical shaker. A triaxial accelerometer array and three readout meters indicate severe lateral motion as you pass through 1,800 Hz. Is the armature (as a unit) moving laterally? Or is there another explanation? Connect your horizontal-sensing accelerometer to the X input of your scope and your vertical-sensing accelerometer to the Y input. Look for one of the patterns of Fig. 18.7. Perhaps cross-talk at 3,600 Hz, 5,400 Hz, or some other multiple of 1,800 Hz is acceptable.

Everything said here deals with sinusoidal waveforms. It is difficult to measure phase difference and to measure frequency ratios unless both signals are "clean." But they may be badly distorted (contain components at multiples or at sub-multiples of f_f, or they may contain "hash"). Use sharply-tuned filters to "clean" them. Automatically tuned "tracking filters" per Sec. 14.7 are very helpful in phase and frequency measurements. Identical motion at all attach points and lack of lateral motion are difficult to observe during random vibration. Perform the checks suggested above (using sinusoidal vibration) before all important random tests. Take the tuning signal from a shaker oscillator, from an engine tachometer, from a photocell monitoring shaft speed, etc. Filters block vibration signals at other frequencies.

Section 19
Test Fixtures

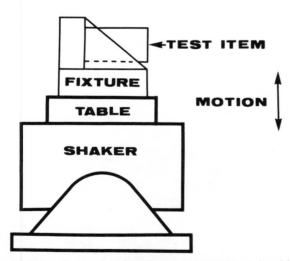

Fig. 19.1 Test fixture couples shaker (or shock test machine) motion to test item.

19.0 Introduction — function of test fixture. The test fixture mechanically couples vibratory energy from a shaker (or shock test machine) to a specimen, per Fig. 19.1. The power amplifier output *should* precisely reproduce its input waveform. The shaker table *should* precisely reproduce that waveform. (But, as discussed in earlier sections, systems do not fully meet these requirements at all f_fs.) The fixture's job is similar: it should precisely reproduce table vibration at the specimen.

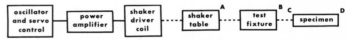

Fig. 19.2 Partly electrical, partly mechanical block diagram of an EM shaker sine test system (replace oscillator with random input for random tests — figure also applies, with some changes, to EH shakers).

19.1 A typical fixture is shown in Fig. 19.3. For testing, it represents a particular location on an auto or truck engine. A control accelerometer ("B" in Fig. 19.2) informs the servo (if a sine test) or spectrum control (if a random test) of fixture motion.

Fig. 19.3 Automobile component attached to fixture representing engine. Circular plate attaches to EM shaker table. Series of intermediate blocks attaches variety of components in various attitudes to shaker.

19.2 Transmissibility revisited. Let Fig. 19.4 represent resonant behavior around a fixture's lowest f_n, say 10,000 Hz. Then our testing to $f_f = 2,000$ Hz will remain in the left portion, will never exceed $f_f/f_n = 0.2$. Tr (per eq. 1.5) will not exceed 1.04. Our vibration controller readily compensates that slight fixture resonant magnification by reducing amplifier power by 4%.

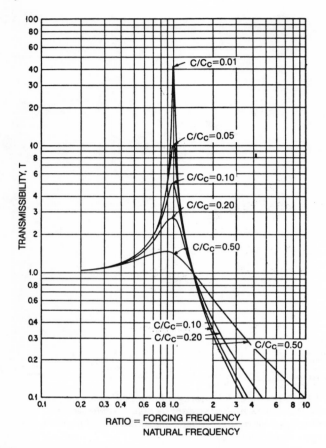

Fig. 19.4 Transmissibility diagram, same as Fig. 1.5, here representing dynamic behavior of fixture.

What if our test item were so large we could not attain $f_n = 10,000$ Hz, say only 3,000 Hz? Then tests to 2,000 Hz would reach $f_f/f_n = 0.67$, and Tr would approach 2. Our controller compensates by dropping shaker power by about 1/2.

What if our best f_n were 2,000 Hz, with tests to 2,000 Hz reaching $f_f/f_n = 1$? Maximum Tr or Q would depend upon fixture damping. Our controller would severely drop shaker power — a task eased by much fixture damping.

What if our test item were so large that our best f_n were only 500 Hz, with ridiculous tests to 2,000 Hz reaching $f_f/f_n = 4$? Tr would drop to some small fraction (fixture isolating test item from shaker). At fixture resonance (500 Hz) our controller must severely drop shaker power — a task eased by much damping. At $1.414 f_n \approx 700$ Hz, Tr = 1. At 2,000 Hz the controller must severely overdrive the shaker, attempting to achieve the specified A at the control accelerometer. The tasks of controller, PA and shaker are greatly eased by much fixture damping, so that Tr does not drop so rapidly.

19.2.1 Components to high test frequencies. Consider fixturing integrated circuits (IC) or other small components, as they will mount in service, as for example on PCBs. They must be removable (without damage) after tests or screening per Sec. 30. An expedient: "pot" hermetically sealed enclosures into non-toxic cetyl alcohol (melts at +130° F — less fire danger than paraffin) cubes. (Experimentally evaluate a cube per Sec. 19.13.) Later, recover the alchohol via distilled water at +150° F. Another idea is shown in Fig. 19.5.

19.2.2 Assemblies to lesser frequencies. An electronic assembly or other "black box" seldom needs $f_u = 2,000$ Hz. Some older "test everything to 2,000 Hz" thinking still exists, but MIL-STD-810D mandates tailoring tests to actual environments. $f_u = 500$ Hz is often high enough, with even lower f_u for large platforms. Further, most internal f_ns (especially within large assemblies) are relatively low. A 1,000 Hz first f_n fixture can be relatively easy to design and will be quite satisfactory to 500 Hz ($f_f/f_n = 0.5$).

19.2.3 Using section of platform as fixture. MIL-STD-810D, page 514.3-33, urges using the maximum possible portion of the platform as your fixture, with actual mounting hardware, so your test item can respond naturally. This conflicts with a tight tolerance on column 4 of Table 19.1.

19.3 Fixture design criteria. Table 19.1 gives design criteria for different sizes of fixtures. Dynamics test personnel specify fixture dynamic requirements to tooling designers and/or procurement personnel. Column 1 describes various sizes, weights and shapes of test items; start here. Column 2 describes allowable Tr peaks; note that columns 2, 3 and 4 become less stringent as you go downward, because the specified f_u *should* be dropping. Sec. 18 and 19.3.2 explain column 3. Column 4 limits differences between the several inputs to a test item; see Sec. 19.3.3.

19.3.1 Fixture resonances can be troublesome when f_u is ridiculously high. (Do not expect quick acceptance of the "tailoring" concept of the 1983 MIL-STD-810D.) But they are

Fig. 19.5 Possible scheme for attaching ICs to shaker. IC leads are pushed into conductive foam pads in a magnesium cube. The cube is successively bolted in three directions to a shaker. Courtesy KII.

less troublesome when you use much hysteretic damping. Consider the alloy (Fig. 19.6) used in some shaker armatures and certain cast magnesium fixtures (Fig. 19.7). Or add damping (such as polyurethane foam, Fig. 19.8).

K1A MAGNESIUM

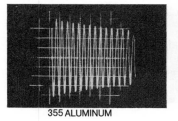

EZ33A MAGNESIUM

AZ81A MAGNESIUM

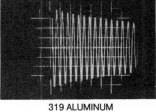

355 ALUMINUM

319 ALUMINUM

Fig. 19.6 Material or hysteretic damping demonstrated by "plucking" cantilevered samples. Signal from small accelerometer shows highest decay (highest damping) on K1A magnesium sample.

Fig. 19.8 Polyurethane foamed in place "saved" this fixture. See Sec. 19.7. Courtesy KII.

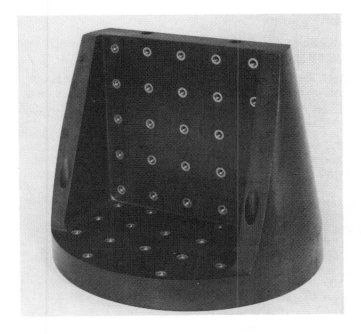

Fig. 19.7 Front and rear views of cast magnesium fixture. Courtesy Avco.

19.3.2 Minimal lateral motion is desired for most testing. This does not represent "the real world," but is easier to instrument. Balanced loading of the shaker removes one cause of armature rocking per Fig. 12.5(b). Static balance your test item + fixture on knife edges. This locates the static cg along the shaker axis. Consider also dynamic balance. If the load of Fig. 19.3 were massive, its resonance could create severe rocking. Try testing two items simultaneously, with a "T" fixture per Fig. 19.9.

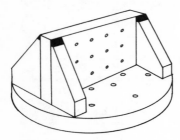

Fig. 19.9 "T" fixture for vibrating items in pairs, thus avoiding rocking moments.

19.3.3 Similar motion at all attach points is wrongly assumed by neophytes. If buying a fixture they may specify that differences be small, per column 4 of Table 19.1. Electronic averaging per Sec. 16.11 permits testing in spite of quality-compromising wide differences.

Table 19.1
DESIGN CRITERIA FOR VARIOUS SIZES OF FIXTURES

(1) Component Description	(2) Allowable Transmissibility Peaks	(3) Allowable Orthogonal Motion	(4) Allowable Variation in Vibratory Input between Test Item Attachment Points
Small components, mechanical, electrical, or electronic, up to cigar-box size and weight up to 5 pounds.	None below 1000 Hz. Above 1000 Hz, a maximum of 3 resonances, limited to 5:1 over 3db bandwidth 100 Hz.	Y and Z motions less than X motion throughout the test range up to 2000 Hz.	±20% allowable up to 1000 Hz. From 1000—2000 Hz, ±50%.
Electrical, electronic, mechanical components in sizes up to a 10-inch cube and weights up to 15 lbs.	None below 1000 Hz. Maximum of 4 peaks above 1000 Hz, 5:1. None to exceed a 3db bandwidth of 100 Hz.	Y and Z motions less than X motion throughout the test range up to 2000 Hz.	±30% up to 1000 Hz. 1000 to 2000 Hz, not exceed 2:1 between any pair of points.
Odd-shaped mechanical components (i.e. large hydraulic actuators and vent relief valves). Electrical equipment (i.e., inverters, telemetering transmitters). Volumes up to 3 ft³, weights 10 to 50 pounds.	None below 800 Hz. Maximum 4 peaks 6:1 over 3db bandwidth 100 Hz, 800—1500 Hz. Maximum 3 peaks 8:1 over 3db bandwidth of 125 Hz, 1500—2000 Hz.	Y and Z less than X up to 1000 Hz. Above 1000 Hz, 2X, except that over a 3db bandwidth of 200 Hz, may be 3X.	±50% up to 1000 Hz. 1000 to 2000 Hz, 2:1, except that over a 3db bandwidth of 200 Hz, input variation may be 2.5:1 between any pair of points.
Larger equipment weighing 50 to 500 pounds, volumes up to 20 ft³.	None below 500 Hz. Maximum 2 peaks 6:1 over 3db bandwidth 125 Hz, 500—1000 Hz. Maximum 3 peaks 8:1 over 3db bandwidth 150 Hz, 1000—2000 Hz.	Y and Z less than X to 500 Hz. 500—1000 Hz, less than 2X, and 1000—2000 Hz, less than 2.5X, except over a 3db bandwidth of 200 Hz, may be 3X.	±50% up to 500 Hz. 500—1000 Hz, 2:1 and 1000—2000 Hz, 2.5:1 except over 3db bandwidth of 200 Hz, variation may be 3:1.
Large equipment over 500 pounds and 24 inches minimum dimension. Note: These fixtures are exceedingly difficult to design. In general, use only with auxiliary hydrostatic bearings.	None below 150 Hz. Maximum 1 peak 3:1, 150—300 Hz; also maximum 3 peaks 5:1 over 3db bandwidth 100 Hz; 300—1000 Hz maximum 5 peaks 10:1 over 3db bandwidth 200 Hz.	Y and Z less than 1.5X up to 300 Hz. Less than 2.5X, 250—2000 Hz except over 3db bandwidth of 100 Hz in range 300—1000 Hz, may be 3:1; also over 3db bandwidth of 150 Hz in range 1000—2000 Hz, may be 4:1.	±50% up to 400 Hz. 400—2000 Hz, 2:1 except over 3db bandwidth of 200 Hz, variation between points may be 3:1.

19.4 Design difficulties. Per Sec. 12.2, shaker armatures are not perfect, even though designs evolve over many years. Expect difficulty when you hurriedly design and build a fixture (at minimum cost and weight). You will never match your shaker armature dynamic behavior and weight.

Simplest: a simple plate. But for three mutually perpendicular directions, it may become the complex structure of Fig. 19.3

or 19.7, with the test item cantilevered. Even with static balance (fixture + specimen), cantilever resonance can generate severe rocking moments and shaker motion per Fig. 12.5(b). Only for small specimens are fixtures so stiff that all resonances are above f_u per Table 19.1, Column (2).

A newcomer's first attempt joins thin aluminum plates with machine screws. Experimental evaluation per Sec. 19.13 shows inadequate stiffness. With his "control" accelerometer at the shaker table center (point A, Fig. 19.2) test item dynamic motion (B or C) at some f_fs is greater and at other f_fs is less than specified. Only at A is the intensity held as specified. Better: place the sensor at B. But other difficulties can appear: (1) At resonance, his servo or random controller dynamic range may be inadequate, over-testing the item. (2) If testing to f_u = 2,000 Hz with a soft fixture having first f_n = say 500 Hz (f_f/f_n = 4), his fixture will isolate (undertest) the test item.

He begins to appreciate why many laboratories go "outside" for fixtures. He may try heavier gauge materials, moving his difficulties to higher f_fs. He may design for machining from a casting or from solid aluminum or magnesium stock. Or he may design for welding. He may add damping (Sec. 19.7).

19.5 Fixture materials. We are seldom concerned with strength or fatigue. Stiffness dictates fixtures so rugged they seldom fail.

Since fixture mass is often critical, aluminum and magnesium are common. For a given stiffness, aluminum is 1/3 heavier than magnesium while steel is five times heavier. Some alloys (see Fig. 19.6) have fair damping. Magnesium bar stock is particularly easy to weld and to machine. The controlling factor for axial f_n is the ratio E/ρ (E is Young's modulus and ρ is density), about the same for most metals. (Beryllium is higher.) Hence the metal selected will not greatly change axial f_n. See also Sec. 20.3.5.

Large (therefore low f_u and f_n) fixtures have been laminated with plastic, wood and/or metals. High damping is their major advantage.

19.6 Fabricating fixtures. After selecting his material, the designer must choose a method for joining the parts.

19.6.1 Casting fixtures has many advantages including ability for complex shapes (such as conical), generous fillets, use of a high-damped alloy, as in Fig. 19.6. Long delivery time usually precludes castings.

19.6.2 Bolts always attach the fixture to the shaker table. And often attach the test article to the fixture. Most engineers overestimate their ability to design with bolts. Mating surfaces often decouple rather than vibrating in unison. Decoupling occurs most commonly at some resonance, when F = MA exceeds bolt preload forces. When using bolts for joining parts of a fixture, consider ideas from Sec. 19.11.

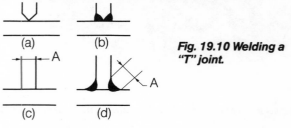

Fig. 19.10 Welding a "T" joint.

19.6.3 Welding is the most popular fabrication method. See Fig. 19.10. Some authorities state that chamfering for full penetration (a) is not needed before welding (b); hold the pieces (c), tack weld each side, then weld a heavy bead on first one side and then the other (d). Repeat if needed, to achieve a fillet width "A" equal to plate thickness.

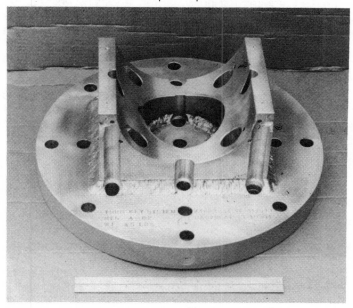

Fig. 19.11 Welded fixture for missile guidance section. Courtesy Turn Key Systems.

19.6.4 Bonding fixtures approaches (at moderate temperatures) the stiffness of welding. Parts are rough sawed, then cleaned. Mix a two-part machineable epoxy such as Devcon; for joining aluminum, one liquid contains aluminum particles. Secure the parts by small screws while the epoxied joints "set up." Two fixtures like Fig. 19.12 were fabricated, one welded and the other bonded; Fig. 19.13 shows that welding was only marginally stiffer. Earlier, curves were plotted while parts were held by screws.

Fig. 19.12 Bonded fixture. Courtesy Martin-Marietta.

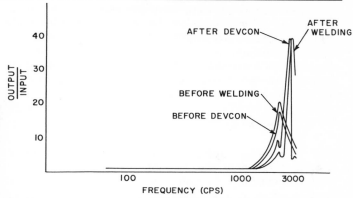

Fig. 19.13 Transimissibility of bonded (as in Fig. 19.12) and welded fixtures. Courtesy Martin-Marietta.

19.7 Damping is helpful for reducing resonant effects. Refer again to Fig. 19.4. When $f_f = f_n$, at resonance, much damping reduces Q (peak Tr); less reduction of electrical power to the shaker is needed. When $f_f \gg f_n$, well above resonance, much damping lessens the Tr loss, requiring less power increase to the shaker. Fixtures may be cast of highly damped magnesium. Other ways to add damping are illustrated in Fig. 19.8 and 19.14.

Fig. 19.14 Laminated fixture joins (with temperature and force) alternating aluminum and epoxy sheets. Segments were sawed and epoxied. (The black masses bring the combined cg close to the shaker axis.) High damping resulted; Q was about 2. Courtesy Burroughs.

On first evaluation, the fixture of Fig. 19.8 resonated at submultiples of f_ns of certain unsupported transverse plates. The fixture had been OK during vertical vibration, but not in horizontal tests. Polyurethane was foamed in place, almost magically converting a poor fixture to a good one.

19.8 Estimating bending frequencies of simple beams and plates utilizes commonly-found formulas. (See Table 19.1, Column 2.) Assembly of such structural elements into a fixture will result in a system with many f_ns. Of these the lowest f_n will determine the f_f range over which the fixture-specimen system can easily be tested. One way to estimate this lowest f_n is by the use of Dunkerley's equation:

$$\frac{1}{f_T^2} = \frac{1}{f_1^2} + \frac{1}{f_2^2} + \frac{1}{f_3^2} + \ldots \frac{1}{f_n^2} \quad (19.1)$$

19.9 Frequency superposition nomograph. The fixture designer often needs easy-to-apply techniques for predicting the f_ns of structural elements that he identifies as beams, gussets, plates, etc. His experimental evaluation (on a shaker) should verify these f_ns, confirming his calculations.

But the various elements, when combined, now exhibit a new frequency, often called the "total" natural frequency, f_T. Estimate f_T by Eq. 19.1 or use the nomograph of Fig. 19.15 to predict f_T. If f_T should be too low, it will predict the effect of design changes.

Eq. 19.1 requires that individual frequencies be well separated, e.g. $f_1 = 100$ Hz and $f_2 = 150$ Hz. The method can still be used if f_ns are very close, but errors approach ±15%. When you analyze the response of a fixture or other structure, first calculate the major responses. Then group these responses via Eq. 19.1 to determine f_T.

Here is a numerical example: Suppose a 2DoF system has computed frequencies $f_1 = 900$ Hz and $f_2 = 700$ Hz. What will be the combined natural frequency f_T? We can insert 900 Hz and 700 Hz into eq. 19.1 and predict that $f_{T1,2}$ will be 550 Hz.

Eq. 19.2 makes the arithmetic a bit simpler.

$$f_{T1,2} = \sqrt{\frac{f_1^2 \cdot f_2^2}{f_1^2 + f_2^2}} \quad (19.2)$$

With a test load attached, a new, lower, frequency should appear. How much lower? You can use Eq. 19.1 or you can rewrite Eq. 19.2 to combine $f_{T1,2}$ with a new f_3. Or again use Fig. 19.15. Assuming that f_3 has been predicted to also be 550 Hz, f_T can be read as 390 Hz.

Likely your no-load experimental evaluation will show a peak at 550 Hz, but not at 700 nor 900 Hz. Your resonance search with load should show the 390 Hz peak. Laboratory management and project personnel must understand Eq. 19.1.

To use Dunkerley's equation, calculate the effective mass M of each fixture structural element and load. Then, *one at a time,* apply each M to the effective stiffness K it sees, and calculate a theoretical f_n by eq. 1.3 or 1.4. Combine these f_ns as shown in the nomograph of Fig. 19.15, to estimate the lowest system f_n. The calculations of Sec. 19.9 assume that all masses move in accordance with the system mode shape.

19.10 Reserve shaker force is helpful. Commonly, a shaker system was purchased for a particular test series. Can we now use it to test heavier loads, or to develop higher A? The test items, fixture and armature may exceed the system's F rating. The designer tries to reduce fixture mass, but it may then be inadequate. Additional money earlier spent for a larger shaker could now be saved in design and test time.

Fixture mass, added to that of the table, is useful. It provides a high impedance to "feedback" energy from the specimen, which might otherwise cause motion distortion and excessive lateral motion. It also eases the shaker control task.

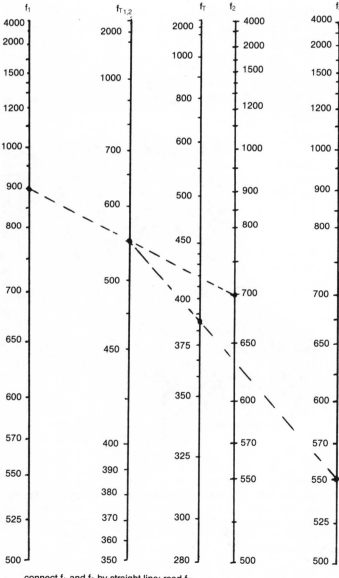

connect f_1 and f_2 by straight line; read $f_{T1,2}$
connect $f_{T1,2}$ and f_3 by straight line; read f_T

Fig. 19.15 Frequency superposition nomograph.

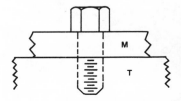

Fig. 19.16 Fixture (or any load, total mass M) bolted to T, a shaker table, a horizontal slip plate or any vibration source. Will M and T have identical vertical motion? horizontal motion?

19.11 Bolting to shaker table. Visualize (Fig. 19.16) a fixture bolted to a shaker table. Vibratory force at this moment is maximum downward. M represents the total mass of fixture + test item divided by the number of bolts (assuming the bolts are equally loaded). The upward force on the bolt is $F = MA$, where A is peak acceleration (of vibration or shock). Unless preload P in the bolt exceeds F, the joint will separate. Aim for $P = 1.1F$ or more.

How do we know initial preload P_I? Most laboratories depend upon torque gages. (Handbooks suggest values.) Unfortunately, these only indicate friction, difficulty in turning the bolt. Friction depends upon dimensional tolerances, lubrication, surface and thread condition. (After a test, don't throw bolts together; threads are thus damaged. Carefully store them in a rack.) Visualize a too-long bolt, bottomed, the mechanic carefully reading torque; unfortunately, $P_I = 0$, due to clearance under the cap. Yes, that *has* happened in many laboratories! With tight control over *all* variables, torque values *can* be used to set initial preload P_I.

Fig. 19.17 Electrical strain-gaged bolt. Courtesy SPS.

Certain bolts indicate their elongation (residual preload P_R as well as P_I) electrically (Fig. 19.17) or visually (Fig. 19.18). Calibrate these in a tensile test machine, elongation vs. force. Better, acoustically measure and record each bolt's length on first tightening, again after tightening others nearby and again later, after "embedment relaxation." Some may need further tightening, and you will *know* which ones.

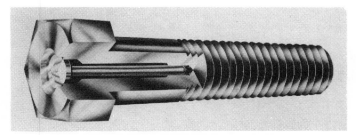

Fig. 19.18 Bolts that visually indicate elongation. Courtesy Modulus.

If we vibrate sinusoidally, we may find a sudden "breakup" of our 'scope pattern. Parts separate at the instant (in each cycle) of maximum tensile force. Upon rejoining, an impact occurs. It can shock-excite our specimen, exceeding the specified A and f_u. This overtesting can cause unwarranted

(and unrecognized) damage. If we test with random, this effect can interfere with equalization (Sec. 25) but may not be recognized.

Drill and tap for at least three bolt diameters. Into soft metals, use a threaded insert. Choose proper bolt shoulder length for the plate being held. Keep this plate relatively thin and use short stiff bolts. Use a heavy steel (machined, not cut) washer to distribute bolt head force to the plate.

Surprisingly, a properly designed bolted connection has a K up to 6 times K_{bolt}. The nearby material (possibly magnesium) in compression acts as a spring, adding to K_{bolt} (probably steel).

The number and size of bolts may be *very* roughly approximated by

$$A = \frac{WL}{E}\left(\frac{f_u}{3.13}\right)^2 \qquad (19.3)$$

This is oversimplified but possibly useful. W is the attached weight in pounds, L is effective bolt length in inches, and E is Young's modulus, 30,000,000 for steel. Divide total area by N. Note that bolt stiffness must rise as f_f^2 in order to avoid decoupling.

For testing to 2,000 Hz, keep the span between bolts under 3 inches. Strips of thin Mylar tape, compressed by bolting, help to prevent localized movement. Alternately, a thin coat of grease or oil between mating surfaces can aid coupling of motion.

19.12 Bolting to auxiliary table. Fig. 19.16 first dealt with forces along the bolt axis, the only proper way to use bolts. Now it deals with forces in shear, as when a fixture is driven horizontally per Sec. 20.3. Newcomers to vibration testing should note that a D of only 50 microinches represents 10g at 2,000 Hz. Clearance between the bolt shoulder and the hole in M thwarts proper joining. Only if there is sufficient friction force (greater than MA) will the motions of M and T be identical. How to get adequate friction? Use tapered pins to transmit shear forces, with bolts used to pull parts together. Best: expanding pins or sleeves, per Sec. 20.3.3.

19.13 Experimentally evaluating fixtures. After you design and fabricate a fixture, test it before use. Dynamic evaluation aids the test engineer, warns him of any fixture resonances below f_u so the test item is not blamed. Ideally, the designer will perform the evaluation himself. He should determine the various f_ns and check these against his earlier calculations. If they differ, he can try to find his errors, furthering his education. Future fixtures will benefit.

Sec. 17.2 gives basic instructions for transmissibility tests and for plotting Tr vs. f_f. Evaluating a fixture is a good example. The criterion is near unity Tr over some f_f range, as in Table 19.1, Column 2, or flat spectral motion at the test article attachment point(s), assuming a flat spectral motion at the shaker table surface.

19.14 Resonant fixtures. Fixture designers have traditionally avoided resonances by keeping fixture f_ns considerably above f_u. Otherwise, per Sec. 19.7, they utilize damping to minimize fixture resonance effects.

But fixture resonance can assist in high-intensity vibration testing, beyond the $F = MA$ limits discussed in Sec. 11.2 (EH shakers — sine), 12.3 (EM shakers — sine) and 23.10 (EM shakers — random). Nankey (Fig. 19.19) described a beam for narrow-band random vibration tests of tantalum capacitors at 100g RMS, centered on 320 Hz. Resonant fixtures have been used to 1,000g; a "family" of beams permits wideband high-g testing.

Fig. 19.19 Resonant beam fixture. After R. Nankey.

Section 20
Auxiliary Devices
for Testing Large Loads

20.0. Vertical motion more convenient. Sec. 20 applies to both EH and EM shakers, which are generally seen with the axis vertical. Why? A horizontal shaker table surface is more convenient. To shake the specimen along three axes, simply rotate specimen and fixture into three attitudes. Horizontal motion is sometimes necessary, per Sec. 20.3, but hazards exist.

Fig. 20.1 Flexing of portions of load overhanging table.

20.1 "Head expander" adds area. Some test items are so large that shakers need more table area. (This discussion also applies to the simple threaded shaft terminations of most EH shakers.) How about an overhanging plate fixture per Fig. 20.1? As a crude guide, if thickness equals overhang, edge flexing $f_n \approx 2{,}000$ Hz; tests above say 1,000 Hz will be compromised. Flexure "bottoming" per Sec. 20.2 will worsen. The added mass of a thick plate will "use up" some shaker force.

Fig. 20.2 Flat plate table extender is supported by air bags. Linear bearings at corners restrain flexing per Fig. 20.1. Courtesy LDS.

Fig. 20.3 "Head expander" on small shaker adds area. Courtesy Shinken.

Consider a "head expander" similar to Fig. 20.3. For a given area, it flexes at a higher f_n and adds less mass.

20.2 Overloaded flexures may "bottom." The weight of a heavy specimen. especially with a "head expander," may "bottom" the flexures, preventing motion. Based on no-load stroke D and the flexure stiffness K, available stroke D_{net} will be:

$$D_{net} = 2\left[\frac{D}{2} - \delta\right] \qquad (20.1)$$

Static deflection $\delta = \dfrac{W}{K} = \dfrac{Mg}{K}$ $\qquad (20.2)$

where W is the weight (or M the mass) added (fixture + test item).

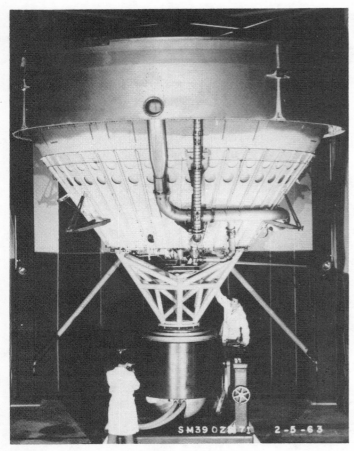

Fig. 20.4 Thin tubing welded into head expander. Polyurethane foam damps tubing resonances. Courtesy Douglas Aircraft.

20.2.1 External vertical support. D_{net} depends upon static deflection from the "no load" position. If you find that D_{net} (with a particular load) is too small, simply counter the gravity force on that load with an external upward force (block-and-tackle, chain blocks or electric hoist). Apply that force through a resilient "bungee" or aircraft shock cord, as in Fig. 20.5.

Fig. 20.5 Resilient "bungee" cord supports weight of test item.

20.2.2 Internal vertical support. A few shakers circulate direct current in the driver coil to support heavy loads. Others have pressurized air bags.

20.3 Supporting heavy loads for horizontal vibration. A *very few* specimens must be tested with a particular side downward. Examples include gyros, units on shock mounts, containers of fluid, etc. To test such specimens, follow Fig. 20.6. Excessive moment (load × distance from shaker table to load cg) may damage flexures or permit armature rubbing.

20.3.1 Auxiliary slip tables. Oil film auxiliary sliding tables per Fig. 20.7 ease horizontal testing of large, heavy structures. Most plates are aluminum or magnesium (maximum stiffness and minimal mass). The smooth, flat bottom surface moves on oil or grease on a block (usually granite).

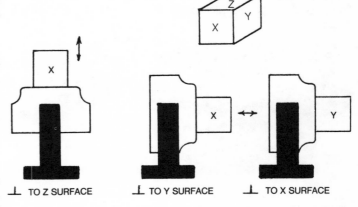

Fig. 20.6 A very few test item categories require shaker horizontal motion.

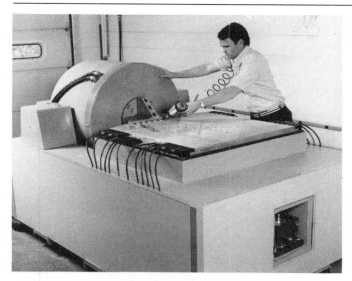

Fig. 20.7 Following tests with shaker axis vertical, armature is being attached to horizontal slip table. Pump (visible at lower right) sends oil via hoses to sliding surface. Courtesy U-D.

Specimens to be tested are bolted to the moving table. The auxiliary table and shaker should mount on a common foundation, which may float on resilient supports.

20.3.2 Slip table no panacea. If the shaker and slip table of Fig. 20.7 had separate bases, three-axis leveling and align-

ment of each unit would be very difficult. With such difficulty you might be tempted to use your slip table for all testing, whether or not required. We object for three reasons:

1. Per Sec. 19.12, bolts (between fixture and slip table) do not hold securely.

2. Horizontal tests, particularly at high f_rs, are better run with the fixture directly attached to the shaker armature, "saving" slip table mass.

3. Axial resonance problems per Sec. 20.3.5 are reduced.

20.3.3 Connecting to the shaker as in Fig. 20.8 "picks up" all table attachment points, better than just one row. This connector joins to a replaceable flat plate by rigid "expanding pins," thus avoiding the "bolts in shear" difficulty of Sec. 19.12.

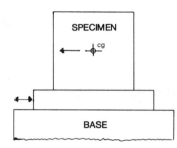

Fig. 20.9 High cg load on slip plate is OK when pulled left. But when driven right, reaction force acts through load cg, producing a couple which can lift the plate. Plate returns with severe shock.

20.3.4 Hydrostatic bearing tables solve the difficulty of Fig. 20.9. Bearings per Fig. 20.10 restrain horizontal motion (can also be used as overhead support or to restrain vertical motion) of severely off-axis cg loads. Oil-film stiffness approx-

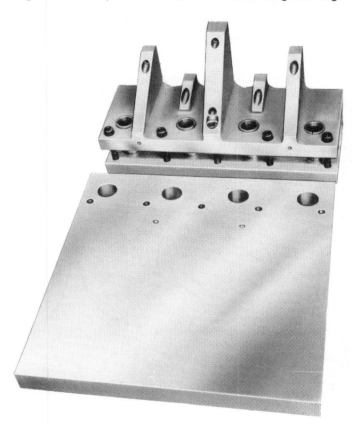

Fig. 20.8 Series of complex magnesium weldments joins any shaker to slip plates. Courtesy KII.

Fig. 20.10 Hydrostatic bearing table is easy to drive horizontally. Outer "T" segments are extremely rigid against pitch while center "V" segment restrains yaw. Courtesy NASA and Team.

imates that of steel. The load attaches to sliding segments driven (through an adapter plate) by a shaker. Recognize that while such restraint expedites testing and protects the shaker, it produces a highly artificial test environment.

Fig. 20.11 Tall satellite (cg up 15 feet) is restrained by six individual hydrostatic bearings. Courtesy Philco-Ford and Team.

Oversized loads may attach via a massive fixture plate to separate T and V bearings, per Fig. 20.11. That plate can be thinner when a slip table is combined with V and T bearings or with less expensive, less restraining journal hydrostatic bearings; Fig. 20.12 shows an application.

20.3.5 Axial resonances. A danger of extending the shaker table per Fig. 20.7 through .12: axial resonance. Fig. 20.13 suggests constant motion close to the shaker but 30dB (roughly 30 times) greater motion at the outboard end. Or if constant motion at the outboard end, -30dB (only 1/30th as much) motion close to the shaker. Only at low f_fs (sine or random) can we keep all points within ±10% of the desired intensity (or for random tests, of the desired spectrum). The use of several accelerometers and electronic averaging per Sec. 16.11 and 27.5 is really only an expedient for control. The difficulty remains.

Frequency f_{axial} depends upon the velocity (compression wave) of sound in the moving system. Let L be the length of that system (shaker armature + auxiliary table) in inches, then

$$f_{axial} \approx \frac{49,000}{L} \tag{20.3}$$

or with L in meters, then

$$f_{axial} \approx \frac{1,250}{L} \tag{20.4}$$

Fig. 20.12 Shipboard antenna receives horizontal vibration atop bearing-line table. Courtesy Plessey and KII.

If we will test to 1,000 Hz and want f_{axial} at least 2,000 Hz (to avoid problems per Fig. 20.13), L must be less than 24.5 in or 0.62 m. With that length, at 2,000 Hz, a series of accelerometers (or test items) along the plate will experience a severe acceleration gradient. Use short plates for high f_f and long plates for low f_f tests. Note: Sec. 20.3.5 applies to long test items, as well.

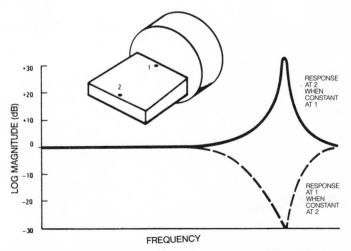

Fig. 20.13 Axial resonance of shaker + auxiliary table limits upper test frequency, sine or random.

One possible solution: restrict the test item and fixture to a small area on the plate. One control accelerometer *may* represent all "input" motions to your test item. That is only possible with a compact test item. So why use an auxiliary plate? Instead, use a compact "L" fixture. You will (1) avoid this resonance and (2) lessen the mass being vibrated.

Do not blindly use auxiliary tables. Within limitations, they are worthwhile accessory devices allowing *low frequency* horizontal vibration testing of large objects.

20.4 Multiple shakers in concert. A large fixture or test platform might be driven by say three shakers. The somewhat unrealistic goal might be non-resonant solid-body platform motion, with all three attachment points experiencing identical motion. The shakers could be powered by a single PA and oscillator, and could develop equal force. Visualize one accelerometer at each attachment point; their signals would be identical.

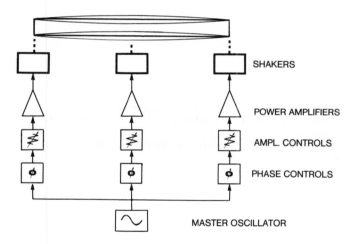

Fig. 20.14 Three shakers (each with own power amplifier, servo and phase control) drive a test platform. Certain platform modes are suppressed.

But at some f_f a resonant mode appears; the platform center motion increases per Fig. 20.14. With a single PA, we cannot regulate individual forces. We need individual PAs. We provide each channel with a servo per Sec. 14.6 or digital control per Sec. 15. However, at the mode shown, the center shaker might receive zero drive. yet the platform center motion might still be too great (force coming from the other two shakers). The platform responds as it "wishes." How to suppress or prevent such a mode? We might add phase control, to reverse the center shaker's force. We might further add electronic cross coupling (not shown) to further reduce power to the other shakers when a critical platform point responds too strongly.

For random vibration testing, a single noise generator could feed an equalizer and PA for each shaker; the several motions will correlate. Most test directors prefer uncorrelated motion, each channel driven by its own noise generator.

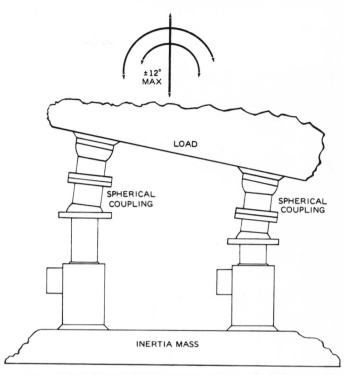

Fig. 20.15 Spherical hydrostatic couplings have zero "slack," and thus are suitable between shaker and load. They swivel, permitting some rotation and some translation. Courtesy Team.

20.5 Spherical couplings permit some out-of-phase motion. Fig. 20.15 shows a load driven by two shakers. Fig. 20.16 shows hydrostatic couplings capable of ±12° motion.

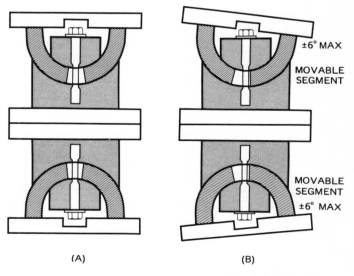

Fig. 20.16 Cutaway of spherical coupling suitable for the applications of Fig. 20.15 and 20.17.

20.6 Multiple shakers — orthogonal motion. Two or three single-axis shakers can drive a load in two or three axes. Refer to Sec. 28.4 re seismic tests and Sec. 30 re stress screening. Fig. 20.17 represents tri-axial EM shaker installations at the US Army's White Sands Proving Ground and Harry Diamond Laboratories, also at a French laboratory. Spherical couplings are needed to permit vertical and north-south motion without damage to the east-west shaker. Likewise to permit vertical and east-west motion without damage to the north-south shaker.

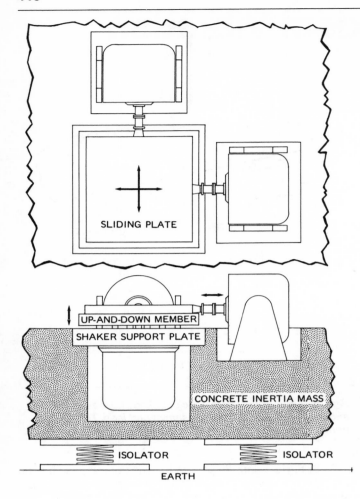

Fig. 20.17 Three-shaker installation for simultaneous vibration in three mutually-perpendicular axes. The vertical-acting shaker is suspended from a steel plate; any horizontal motion of its load is prevented by linear hydrostatic bearings (not shown). Both horizontal-acting shakers drive a two-axis slip plate.

Section 21
Introduction to Random Vibration

21.0 Where is random vibration found? Sec. 21-27 deal with random vibration measurement, analysis and testing. Aerospace activities have longest used broadband random vibration testing to represent flight.

1950s oscillograph records representing random vibration of early missiles and jet aircraft were markedly different from familiar complex vibrations on propeller-driven aircraft. Visual examination was useless. Spectral analysis showed a wide, continuous span of frequencies. Sources (see Fig. 21.1):

1. Turbulence around high-velocity exhausts, at lift-off of a ground-launched vehicle. Random sound causes the vehicle skin to vibrate randomly. Then, as with turbulent air flow, the skin vibrates randomly. Two paths cause interior component motion: (a) via the vehicle structure (mechanical coupling) and/or (b) via the skin re-radiating intense sound. Somewhat similar exhaust sounds emanate from turbojet and turbofan engines, accompanied by complex sounds from turbine and fan blades.

2. Turbulent air flow along the skin of missiles and high-performance aircraft, most severe during transonic flight.

3. Very low frequency engine thrust reaction; this may affect human occupants and may create liquid fuel flow problems.

Whereas aerospace random vibration frequencies reach thousands of hertz, earthquake (seismic) random vibration is below 50 Hz. We find some random sound and vibration in turbulent flow of liquids or gasses. Also in road inputs to automobiles and trucks. Also in wave inputs to ships. Also, random vibration aids production stress screening of electronic subassemblies and assemblies (whether destined for a rugged or a benign environment).

21.1 A quick review of sine vibration. Spend a few minutes reviewing Sec. 2. Sinusoidal vibration is extremely rare, occurring only in calibration laboratories.

The propulsion systems of propeller aircraft, cars, trucks and ships create "complex sinusoidal" vibration. Here (1) the waveform is cyclic and repetitive; (2) energy exists at definite, discrete frequencies. Review Sec. 7.

21.2 "Hash" is a term that covers nonrepetitive waveforms, as in Fig. 21.2, usually with most energy at high frequencies,

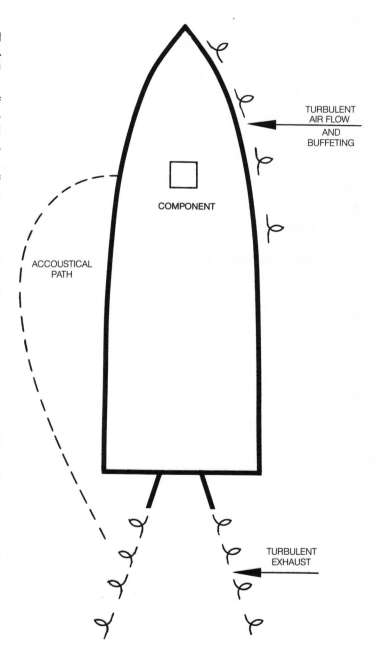

Fig. 21.1 Sources of random vibration input to delicate on-board components.

caused by parts of a vibrating system striking (as when bolts do not have enough preload or by a loose or poorly connected accelerometer). Hash somewhat resembles random vibration.

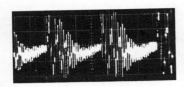

Fig. 21.2 "Hashy" waveforms from accelerometer on shaker table during attempted sine vibration test.

21.3 Common random sounds. When you listen to the sound of a rocket engine (or, on a smaller scale, a blow torch) you hear a distinctive sound: audio energy that is composed of a wide band, a wide spectrum, of frequencies. Other random sounds: road/tire noise transmitted through your car's suspension; turbulent air noise through your car's windows when you drive fast; a waterfall; bacon frying; your TV set (on a "blank" channel).

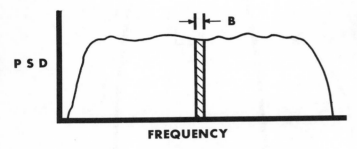

Fig. 21.3 Understanding broad-spectrum random vibration is eased by studying a frequency "slice" of that spectrum within bandwidth B.

21.4 Looking at random signals. An experiment per Fig. 21.4 will aid understanding. The noise generator and band-pass filter should cover say 5-5,000 Hz. Bypass the filter; noise signals of all frequencies reach the 'scope and your ears. Insert the filter and adjust it to pass frequencies from

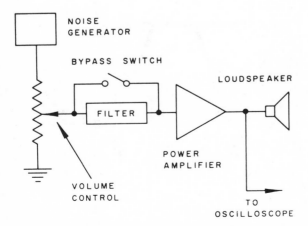

Fig. 21.4 Block diagram of equipment for demonstrating properties of random electrical signals.

20-2,000 Hz (typical frequency range for some random vibration test specifications); your 'scope picture may resemble Fig. 21.5.

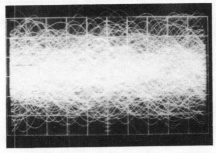

Fig. 21.5 Oscillo—scope time history of 20-2,000 Hz random signal. While shutter was open, beam traced out many complex patterns, all different.

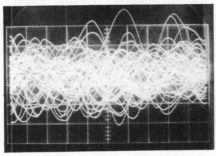

Fig. 21.6 Oscillo-scope time history of 20-500 Hz random signal. While shutter was open, beam traced out many complex patterns, all different.

Readjust the filter to pass only 20-500 Hz, giving something like Fig. 21.6; now only frequencies to 500 Hz reach the 'scope. Note the different sound, now resembling a rocket engine. We cannot predict the instantaneous signal magnitude; hence the term "random." But we can observe that high instantaneous peaks occur least often. Most of the time, the instantaneous voltage is close to zero. We will discuss this quantitatively in Sec. 22.10.

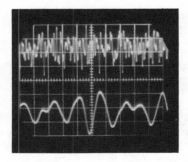

Fig. 21.7 Oscillo-scope "single sweep" time history. Upper trace 20-2,000 Hz. Lower trace approximately 1/3 octave wide, centered at 100 Hz.

Fig. 21.7 was taken with the 'scope in a slow "single sweep" mode. The filter for the lower trace was peaked at 100 Hz, with about 30 Hz bandwidth B. That trace resembles a 100 Hz sine wave, except that instantaneous amplitude varies unpredictably. Can you hear the amplitude-varying 100 Hz signal? If possible, slowly sweep a narrow filter up to 5,000 Hz; observe waveforms and sounds.

Now you should understand that random motion (represented electrically here) is *not* cyclic and *not* repetitive in detail, and that energy exists over a continuous band of frequencies.

Fig. 21.5 and 21.6 and the upper trace of Fig. 21.7 might represent vibration signals from an accelerometer; or force signals from a force gage; or pressure signals from a microphone or pressure pickup. All might represent broad-spectrum dynamic inputs to a structure. If that structure had a single resonant peak around 100 Hz, its response might resemble the lower trace of Fig. 21.7.

21.5 Need for random testing. Current military and other test specifications call for random vibration of structures (including electronic parts). They respond differently to random forces than to sine forces. Test vibration should at least approximate service vibration.

21.6 No equivalence to sine vibration. Fig. 21.8 shows a device for demonstrating the virtues of random testing. It could represent an array of PCBs, each with its own f_ns. Use a wood or plastic block 1 inch thick and perhaps 4 inches square. Two saw cuts at 45° clamp two metal reeds having markedly-different f_ns, say 17 and 36 Hz. With a shaker, vibrate the block sinusoidally in the direction shown. One *or* the other reed, at resonance, has large D. They never touch, no matter how hard you drive the shaker. (If you drive too hard, or if you dwell at resonance too long, a beam will fatigue. But that is not our goal here.)

Now drive the shaker randomly. Each reed's motion will resemble Fig. 22.8, one at 17 Hz and the other at 36 Hz. They touch occasionally, at moments when both have large D. Similarly, PCBs strike in service and during random tests, though not during sine tests.

Some people fear and oppose random vibration testing because they don't understand it. This device demonstrates its merits. You now understand why random is required for missile and aircraft testing, also for production stress screening per Sec. 30. Also for seismic simulation per Sec. 28. There is no "equivalent" random test.

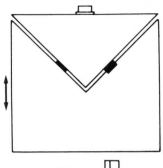

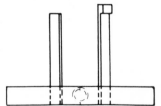

Fig. 21.8 Random vibration "demonstrator".

Section 22
Measurement and Analysis
of Random Vibration

22.0 Introduction. Sec. 21's introductory discussion was qualitative and descriptive. Now we will quantize and express random vibration intensity as Acceleration Spectral Density (ASD) or Power Spectral Density (PSD), in g²/Hz. Also define more fully what the word "random" means in dynamics. Let us imagine a quantitative experiment, pretending that we are measuring a random vibration. Refer to Lord Kelvin's dictum, Sec. 2.0.

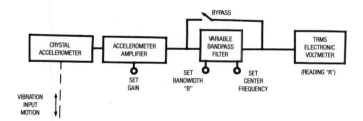

Fig. 22.1 Analog instruments for spectral analysis.

22.1 Suitable instruments. See Fig. 22.1. We need an accelerometer, suitable amplifier, variable band-pass filter and a *true* RMS voltmeter such as the Hewlett-Packard 3400A, the B & K 2417, or the Ballantine 320.

22.2 Calibration with sinusoidal motion. First, we will calibrate our instruments, much as for sinusoidal motion, at say $f_f = 1,000$ Hz. Temporarily bypass the filter.

"USA" UNIT EXAMPLE
Vibrate the accelerometer sinusoidally at 1g RMS (equals 1.414g peak) on a calibration shaker. Adjust accelerometer amplifier gain to give a voltmeter reading of exactly 1 volt RMS. Thus the meter reading is direct: 1

volt RMS indicates 1g RMS acceleration.

INTERNATIONAL SYSTEM EXAMPLE
Follow the same procedure, except shake at 10 m/s² RMS (equals 14.14 m/s² peak). 9.81 m/s² is more precise, but for simplicity let's use 1g = 10 m/s². Now 1 volt RMS indicates 10 m/s² RMS.

Next, vary f_f over a very wide range, say 1 to 10,000 Hz, to insure that system sensitivity remains constant. Return f_f to 1,000 Hz. (See Sec. 9.7.3 re calibration with random forcing.)

22.3 Effect of filter on sinusoidal signals. Now insert the filter, adjust its bandwidth to 160 Hz, then locate its f_o at 1,000 Hz. The voltmeter reads 1 volt RMS because our motion is between 920 and 1,080 Hz (filter has no effect within the pass band.) The meter reads zero if vibration is below 920 or above 1,080 Hz (idealized filter).

22.4 Effect of filter bandwidth on random signals. Now we are ready to measure a vibration which in Sec. 22.7 is found to be broad-band "white" random. We attach our accelerometer to this idealized vibration source, then set the filter's f_o and its B to permit measuring the random vibration 920-1,080 Hz. Our voltmeter reading is 4 volts RMS. We deduce that the vibration's RMS value (920-1,080 Hz) is exactly 4g (40 m/s²).

However, suppose that we had not chosen to investigate 920-1,080 Hz with a 160 Hz filter but rather 980-1,020 Hz with a 40 Hz filter. We would have gotten a lesser reading. How much less? Electronic theory states that the filter's output voltage is proportional to $\sqrt{B}$. Since we have dropped B by 4:1, our reading will drop by 2:1, to 2 volts RMS, representing 2g (equals 20 m/s²) RMS acceleration. Or, if we had chosen to investigate 995-1,005 Hz with a 10 Hz filter, our meter would read 1 volt RMS, indicating 1g (equals 10 m/s²) RMS acceleration. Has the vibration lessened? No. We've simply measured less of that total. (We'll measure the total in Sec. 22.7.)

22.5 Various analyzers differ in bandwidth. Our instrumentation is essentially that found in older analog electronic analyzers still used in some acoustical and vibration work. Fig. 22.2 suggests measuring vibration in different parts of the spectrum. Among the various analyzers, differences exist as to filter B. Thus different analyzers could disagree. Three investigators could obtain three different answers, unless all happen to use the same B.

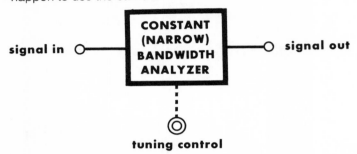

Fig. 22.2 Analog filter can be swept or stepped across a frequency range of interest.

22.6 Spectral density. Rather than describe the vibration around 1,000 Hz in terms of RMS readings, let us square these into "mean square" readings. Finally, let us divide by the B we've used each time. Our result: Acceleration Spectral Density (ASD) or Power Spectral Density (PSD). Calculations are tabulated in Table 22.1.

Table 22.1

"USA" UNIT EXAMPLE

Bandwidth Hz	RMS g	MS g²	ASD g²/Hz
160	4	16	0.1
40	2	4	0.1
10	1	1	0.1

INTERNATIONAL SYSTEM EXAMPLE

Bandwidth Hz	RMS m/s²	MS m²/s⁴	ASD m²/s³
160	40	1600	10
40	20	400	10
10	10	100	10

Expressing ASD in g²/Hz or m²/s³ describes random vibration power (analogous to volts²)in a frequency window. No matter what our B, we get the same answer. We have "divided out" B effects. We "normalized" our measurements—we pretended that everyone uses 1 Hz B.

22.7 Investigating the entire spectrum. How severe is vibration elsewhere? We tune our analyzer from zero to ∞ frequency. We vary f_o, keeping B constant. Most analyzers plot ASD vertically, while varying f_o moves the display horizontally. The "real world" result might resemble Fig. 22.3.

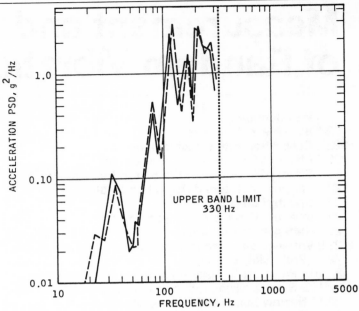

Fig. 22.3 Spectra of liftoff vibrations (two flights). Telemetry limited analysis to 330 Hz.

When random vibration ASD is essentially flat over a wide, continuous band, we may call it "white," an analogy with "white" light—equal intensity at various wavelengths. However, note that vibratory energy is zero at any single frequency (zero B).

Truly "white" random vibration never occurs on real structures. Structural resonances and antiresonances cause ASD peaks and valleys. However, it is a simple concept—one that for years appealed to test specification writers.

22.8 RMS g from root area. Let us now calculate the RMS acceleration inputs to three imaginary mechanical resonators, having respective bandwidths 10, 40 and 160 Hz. Assume that input ASD is constant 0.1 g²/Hz (10 m²/s³). To get the mean-square inputs, we calculate the area under each ASD curve; for simple rectangular spectra, multiply height × width. $\sqrt{\text{Area}}$ gives the RMS acceleration. See Table 22.2.

Table 22.2

"USA" UNIT EXAMPLE

ASD g²/Hz	Width Hz	Area g²	$\sqrt{\text{Area}}$ RMS g
0.1	10	1	1
0.1	40	4	2
0.1	160	16	4
0.1	2,000	200	14+

INTERNATIONAL SYSTEM EXAMPLE

ASD m²/s³	Width Hz	Area m²/s⁴	$\sqrt{\text{Area}}$ m/s² RMS
10	10	100	10
10	40	400	20
10	160	1,600	40
10	2,000	20,000	141

Note that lines 1, 2 and 3 in the final column in each example equals the original readings in Sec. 22.4. Note that sharper resonators receive less input than does a broad resonator, though to predict their response we multiply that input by a larger factor. Line 4 column 4 in each example would result if the random vibration were "white" over a band 2,000 Hz wide and the signal was not filtered. Measurements of overall, wide-band random acceleration are needed in random vibration testing, per Sec. 23.7 and 23.10, so we do not exceed shaker force ratings.

Note that the RMS acceleration is not a complete description. We need information about the time domain per Sec. 22.10. We need information about the frequency domain per Sec. 22.7.

22.9 Variance and standard deviation. "Mean square acceleration" is represented by the area under the ASD vs. frequency graph; statisticians use a different but analogous term: the *variance*. The symbol is σ^2. "Root mean square acceleration" is represented by the square root of that area; statisticians use a different but analogous term (square root of the variance): the *standard deviation*, or σ.

22.10 Probability density. In much of Sec. 22, we discuss random vibration in terms of frequency, both narrow-band and wide-band random vibration, in the *frequency domain*. Let us now examine the *time domain*. We should assess the relative likelihood (or probability of occurrence) of various amplitudes. Both descriptions (frequency domain and time domain) are necessary. Unfortunately, many vibration test laboratories ignore the latter. They should seriously consider analog probability density analyzers or should use that feature of a digital shaker control.

Since all frequencies in a band are present, "broad band" random vibration is harder to describe than is sinusoidal vibration. A sine wave is completely described if we know maximum value and frequency. In Sec. 2.4 we expressed instantaneous sinusoidal acceleration:

$$\ddot{x} = -2\pi^2 f^2 D \sin 2\pi f t \qquad (22.1)$$

The instantaneous value of random acceleration cannot be predicted. However, its statistical properties can be stated (by rather complex math). How much of the total time is the instantaneous $\ddot{x}$ within certain limits? Per Sec. 2.7, with sinusoidal vibration there is a constant 1.000:0.707 peak:RMS ratio. There is no comparable relationship in random vibration. Some peaks will be greater than 1σ, the RMS level; some will be less than 1σ.

Per Sec. 21.4, severe peaks are infrequent. There are as many positive as negative peaks. With much observation, we would discover Fig. 22.4. Intensity is given vertically on a "normalized" scale where 1σ = the RMS level, 2σ = twice the RMS level, etc. The instantaneous $\ddot{x}$ will be between $\pm1\sigma$ 68% of the time; between $\pm2\sigma$ 95% of the time; between $\pm3\sigma$ over 99% of the time. Only 0.3% of the time, or 3 milliseconds per second, will peaks exceed 3σ.

Describing a 6g RMS vibration test may clarify this point. The instantaneous acceleration will be 6g or less about 68% of the time; 12g or less about 95% of the time; 18g or less over 99% of the time.

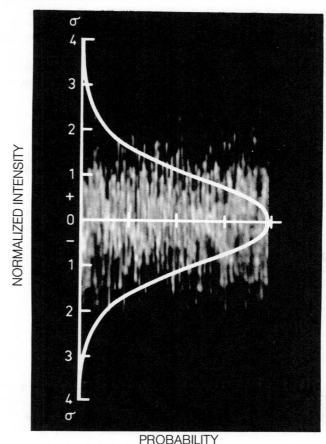

PROBABILITY
Fig. 22.4 Probability density graph (for "Gaussian" or "normal" distribution) plotted over oscilloscope time history (several sweeps) of broadband random vibration. Smallest amplitudes have highest probability (to the right).

Peaks greater than 3σ are very rare. In practical vehicles such as rockets, peaks much over 3σ would destroy the vehicle. With a 6g RMS acceleration level, very few peaks will exceed, say, 18g. Only 0.3% of the time, 3 milliseconds per second, will $\ddot{x}$ be greater than 18g. See Sec. 27.4 concerning peak "clipping" during random vibration testing.

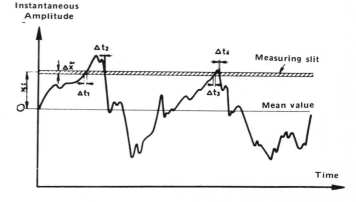

Fig. 22.5 An analog concept of obtaining Fig. 22.4. $\ddot{x}$ appears on a 'scope. Visualize figure width reduced to zero.

22.10.1 Probability density analysis. Fig. 22.5 suggests how to obtain the bell-shaped line of Fig. 22.4. Reduce display width to zero, then slowly sweep a measuring slit upward, through the amplitude range. The 'scope beam occasionally flashes through the slit, most often when the slit is close to zero, the mean value. Monitor the flashes with a photocell, integrating to move a plotter horizontally to obtain a plot similar to Fig. 22.4, with maximum probability at zero acceleration.

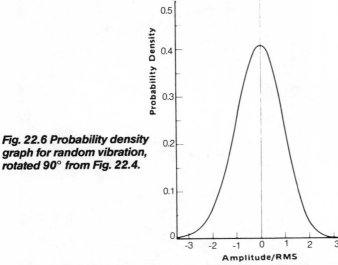

Fig. 22.6 Probability density graph for random vibration, rotated 90° from Fig. 22.4.

Is your shaker probability density "Gaussian" as specified? Electrical faults in the test console (such as unwanted periodic components) or power amplifier, also certain mechanical conditions on shaker or test items, can skew probability density. Quality Control should insist upon this analysis.

22.10.2 Sine vibration probability density. Suppose that the "moving slit" analog analysis of Sec. 22.10.1 yielded the graph of Fig. 22.7. What waveform would appear in the time domain on a 'scope? $\ddot{x}$ never exceeds $\pm 1.414\sigma$. It spends

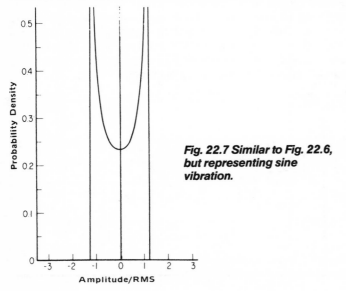

Fig. 22.7 Similar to Fig. 22.6, but representing sine vibration.

most time at or close to its $\pm 1.414\sigma$ peaks, and least time at or close to zero. It must be a sine wave.

22.11 Narrow band random response. Suppose that we shift our accelerometer from a broad-band input random vibration location (such as close to a rocket engine), to a narrow-band response random vibration location, on a sub-structure having a high "Q" resonant amplification at, say, 100 Hz. This sub-structure ignores most of the f_rs in the broad-band vibration input. It only responds to those f_rs close to 100 Hz, and ignores others. Fig. 22.8 shows a "quick look" at a 'scope connected to our accelerometer. The waveform resembles a sine wave whose amplitude varies in a random, unpredictable manner. An ordinary voltmeter measuring this vibration would fluctuate violently.

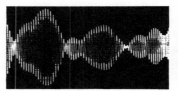

Fig. 22.8 An oscilloscope "quick look" at a narrow band random time history.

The occasional response peaks to occasional input peaks near 100 Hz will eventually fatigue a structure or damage parts mounted on the structure.

22.12 Random vibration tests are gentle. Consider first the frequency domain. Structures respond strongly to sine vibration at or close to their resonant frequencies. With sinusoidal vibration input at an f_n, a structure's response will equal input RMS vibration level × mechanical "Q" factor. There is much less build-up with random vibration. Much smaller responses are noted if the same RMS vibration input is wide-band random, because only a small part of the input is close to f_n.

Consider now the time domain. Large resonant buildups with sinusoidal vibration depend upon regularity of amplitude and phase in the input. Visualize pushing a child's swing. If phase is not just right, the response amplitude will be small. In random vibration, we find only a few consecutive excitation cycles occur close to an f_n. Responses die out rather quickly, as in Fig. 22.8.

22.13 Why test with random? The main value of random vibration is that it induces complex responses in structures and mechanisms by *simultaneously* exciting many resonances.

22.14 Transportation environments can to some extent be simulated with shakers. Some people still use mechanical shakers (Sec. 10), though single frequency vibration does not represent the "real world." EH shakers (Sec. 11) seem best: long stroke, high force at low f_rs, and capable of random vibration. Military contractors should read Method 514.3 of MIL-STD-810D. The Standard differentiates between tests on secured cargo, restrained cargo and loose cargo. Only use "default" spectra if you cannot determine the actual environment. It is much better (usually gentler) to tailor your tests to the environment actually found on the appropriate truck, rail car, transport aircraft, etc.

Section 23
Random Vibration Test Specifications

23.0 Introduction. Sec. 23 follows Sec. 16 on sine testing, but now we move into the 1980s with random vibration testing. We have encountered random vibration in Sec. 21 and 22. Sec. 23.1 uses a simple 1955 spectrum for an example. MIL-STD-810D, dated July 19, 1983, requires more complex "tailoring" to match field conditions.

Section I-3.2.4 of -810D (Method 514.3) lists the major sources of vibration aboard propeller aircraft (random on random or ROR), while I-3.2.5 lists the major sources on jet aircraft and certain missiles (broader spectra). I-3.2.6 deals with helicopters (sine on random or SOR). Other sections deal with external stores. Method 514.3 describes random vibration tests for these applications, also low frequency random tests for ship applications. Sec. I-3.2.12 describes "minimum integrity" tests, unrelated to field environments, for items which don't fit other categories.

Section I-4 points out that the use of RMS g values alone is not valid, that the test spectrum must be shaped. We will discuss equalization in Sec. 25.0.

I-4 also encourages the specifying and control of vibration responses (as opposed to the traditional "input" control described in most of this text) to achieve close correlation between laboratory and service conditions. Your authors heartily concur. We also concur with suggestions on multiaxis excitation. And with ideas concerning early engineering development testing.

Appendix A of Method 514.3, MIL-STD-810D, suggests methods for developing laboratory test specifications.

Generating specifications is beyond the scope of this text. Readers may wish to consult SVM-8, also Bendat & Piersol, Sec. 35.2. In Sec. 22 we cursorily examined the kinds of instrument used for frequency and time domain measurements. One goal: to describe an environment, for the purpose of developing an environmental test. Just as there is much

measurement uncertainty, so is there much uncertainty in developing tests. Along with uncertainty comes risk: the risk of undertesting and "passing" a unit that later fails in service vs. the risk of overtesting and "failing" a unit that would have worked well in service.

Obtaining the environmental description involves properly mounting sensors, correctly connecting cables and signal conditioners, correctly using readout instruments (meters, analyzers, plotters). In many environmental investigations, a further complication: telemetry links for remote measurements. We acquire much data, often in a very brief time; this necessitates computers for data acquisition and processing.

You may be asked to "notch" a spectrum (to reduce ASD in a particular f_f region) to avoid a possible overtest. Notching can modify an input from a very stiff fixture, make the input resemble say a soft aircraft structure. Test item resonant

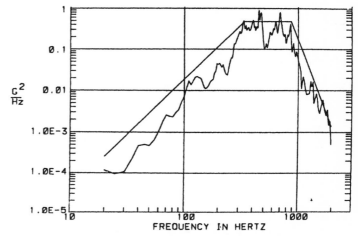

Fig. 23.1 An in-service random vibration spectrum (jagged trace) + typical oversimplified specification, pre-MIL-STD-810D.

responses should not greatly exceed those measured in flight. (We must know what responses are "normal.") Alternately, a sensor can be located within the test item, with its signal "taking over" control (close to resonance) by selection per Sec. 16.11.

23.1 Root area gives RMS acceleration. Sec. 22.6 discussed the frequency-domain terms ASD/PSD, possibly graphed as in Fig. 23.1. To get the area under the curve, we could mathematically integrate the curve, if we had an equation for ASD vs. f_f. Or we could use a planimeter, if our graph were plotted on linear scales. Or, more practically, knowing ASD at certain f_fs and the ASD slopes in between, we can calculate each segment's area, add, then take the square root of that total area. Digital computers (per Sec. 26) readily implement the latter.

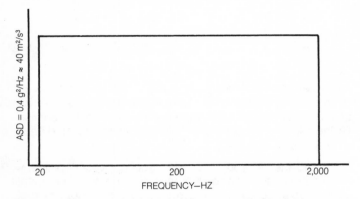

Fig. 23.2 The earliest (known to the authors - 1955) random vibration test specification. Infinitely-sharp rolloff at 20 and 2,000 Hz could not be attained. Area calculations are simple; see text.

23.2 Rectangular spectrum example. As a very simple example, consider Fig. 23.2, with ASD = 0.4 g²/Hz ≈ 40 m²/s³ representative of 1955 thinking and contracts. That oversimplification gives us an easy example. The total area is

"USA" UNIT EXAMPLE

$$\text{height} \times \text{width} = \text{area}$$

$$0.4 \text{ g}^2/\text{Hz} \times 1{,}980 \text{ Hz} = 792 \text{ g}^2.$$

$$\sqrt{792 \text{ g}^2} = 28.1\text{g RMS}.$$

INTERNATIONAL SYSTEM EXAMPLE

$$\text{height} \times \text{width} = \text{area}$$

$$40 \text{ m}^2/\text{s}^3 \times 1{,}980 \text{ Hz} = 79{,}200 \text{ m}^2/\text{s}^4$$

$$\sqrt{79{,}200 \text{ m}^2/\text{s}^4} = 281 \text{ m/s}^2.$$

Per Sec. 25 or 26, the operator compensates or equalizes his shaker to achieve the desired ASD graph. In this example, 28.1g (or 281 m/s²) is the expected reading upon the TRMS voltmeter monitoring overall, broad-band acceleration. Some cardboard or plastic "vibration computers" can approximate the above, although not on practical spectra.

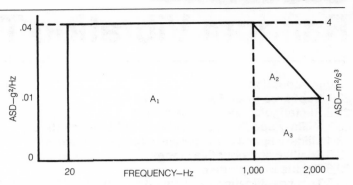

Fig. 23.3 Minimum integrity test for aircraft external store equipment, from Fig. 514.3-36 of MIL-STD-810D.

23.3 Military spectrum example, straight-line method. Suppose that we must comply with Fig. 23.3. What will our TRMS meter read? Note the -6dB/octave slope. Per Eq. 2.16, −6 dB indicates a power reduction of 4:1. In the 1,000 to 2,000 Hz octave, 0.04 g²/Hz (≈ 4 m²/s³) ASD must fall to 0.01 g²/Hz (≈ 1 g²/Hz)

First we divide the total area into three convenient sub-areas: A_1, A_2 and A_3. A_1 is easy:

$$\text{height } 0.04 \text{ g}^2/\text{Hz} \times 980 \text{ Hz} = 39.2 \text{ g}^2$$
$$\approx 4 \text{ m}^2/\text{s}^3 \times 980 \text{ Hz} \approx 3{,}920 \text{ m}^2/\text{s}^4.$$

Likewise, $A_3 = 0.01 \text{ g}^2/\text{Hz} \times 1{,}000 \text{ Hz} = 10 \text{ g}^2$
$$\approx 1 \text{ m}^2/\text{s}^3 \times 1{,}000 \text{ Hz} \approx 1{,}000 \text{ m}^2/\text{s}^4.$$

Triangular area $A_2 = 1/2 \text{ height} \times \text{width}$

$$= 0.5 \times .03 \text{ g}^2/\text{Hz} \times 1{,}000 \text{ Hz} = 15 \text{ g}^2$$
$$\approx 0.5 \times 3 \text{ m}^2/\text{s}^3 \times 1{,}000 \text{ Hz} \approx 1{,}500 \text{ m}^2/\text{s}^4.$$

Now add the three areas to get the total area = 64.2 g² ≈ 6,420 m²/s⁴, and finally take the square root 8g RMS ≈ 80 m/s² RMS.

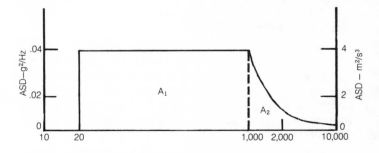

Fig. 23.4 Same spectrum as Fig. 23.3, but plotted on linear graph paper.

23.4. Military spectrum example, exponential method. Fig. 23.3 is misleading. The sloping line is straight only on log-log graph paper. On linear graph paper, Fig. 23.4, it becomes an exponential. Finding the area requires slightly more complex mathematics. Area 1 is the same as before, 39.2g²/Hz ≈ 3,920 m²/s⁴. To calculate A_2 (which in Fig. 23.3 was $A_2 + A_3$), we use equation 23.1.

$$A_2 = \frac{Pf_l}{Z_2}\left[1 - \frac{1}{\left(\frac{f_u}{f_l}\right)^{Z_2}}\right] \quad (23.1)$$

P is the ASD, 0.04 g²/Hz ≈ 4 m²/s³ at f_l = 1,000 Hz. Z_2 is 1, per Sec. 23.5, based on the rolloff rate, here -6 dB/octave. f_u is here 2,000 Hz. Then, per eq. 23.1, A_2 = 20 g² ≈ 2,000 m²/s⁴, and the total area is 59.2 g² ≈ 5,920 m²/s⁴. Taking the square root gives us the RMS acceleration: 7.7g ≈ 77 m²/s. These are better approximations than 8 g ≈ 80 m/s² per Sec. 23.3, and this is the method used by computers per line 9, Fig. 26.3.

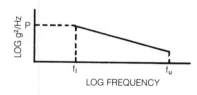

Fig. 23.5 Identification of several terms in Eq. 23.1.

23.5 Positive and negative slopes. Fig. 23.5 identifies several terms of eq. 23.1. For most negative slopes, Z_2 is calculated by

$$Z_2 = \frac{R}{3} - 1 \quad (23.2)$$

In our example, with R (the rolloff rate) -6 dB/oct, Z = 1. Had R been -3 dB/octave, we would have used

$$R = 2.3 Pf_l \log\frac{f_u}{f_l} \quad (23.3)$$

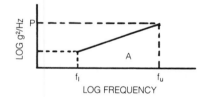

Fig. 23.6 Identification of several terms in Eq. 23.6.

If instead the slope had been positive, Fig. 23.6 identifies various terms. Z_1 would be calculated by

$$Z_1 = \frac{R}{3} + 1 \quad (23.4)$$

Then

$$A = \frac{Pf_u}{Z_1}\left[1 - \left(\frac{f_l}{f_u}\right)^{Z_1}\right] \quad (23.5)$$

would be used to calculate the area under the rising portion.

23.6 Engineering studies. Until recently, we seldom used random vibration for engineering studies of products and structures. Sinusoidal vibration minimized equipment needed for identifying f_n, C/C_c, Tr, etc. but consumed much time.

Random vibration simultaneously excites all resonances, saving much time. Digital computers aid greatly in processing and studying the vast amount of data from such investigations. Two shakers (uncorrelated signals) are better than one; all resonances are more likely to be excited.

23.7 Test specifications - frequency domain. Random test specifications dictate ASD, either by descriptive words or by a graph resembling Fig. 23.3. Some give the total RMS acceleration (needed to assure adequate shaker RMS force). If not, calculate this per Sec. 23.3 or 23.4. Or let your shaker's digital computer (see Line 9 of Fig. 26.3) calculate this.

The test monitor should verify that ASD tolerances are not exceeded. Further, per Sec. 13.6 and 27.4, that no unwanted vibration exists above f_u. To meet specifications resembling Fig. 23.3, a very sharp cutoff 2,000 Hz filter is needed. However, many systems allow the ASD to roll off at say -6 dB/octave; specimen resonances above 2kHz could be improperly excited. Examine the spectrum to, say, 10kHz. If necessary, obtain a separate RTA per Sec. 7.10 for this examination. Other sources of unwanted excitation above f_u: (1) power amplifier distortion, per Sec. 13.7, and (2) accidental (or deliberate per Sec. 23.8 and 27.4) signal clipping.

ASD tolerances in the 1950s were typically ±3 dB. Per Eq. 2.16, those tolerances could be stated as +100%, -50%. Tolerances today (based upon digital control) are more commonly ±1.5 dB, about +41%, -29%.

23.8 Test specifications - time domain. Most random test specifications dictate "Gaussian" or "normal" amplitude probability density, discussed in Sec. 22.10. The test monitor should verify this. Many specifications permit ±3σ clipping per Sec. 27.4.

23.9 Sinusoidal equivalence. Early (1955) random test specifications startled many laboratories. Many had sinusoidal equipment, often powered by alternators not adaptable to random testing. For labs incapable of random testing, a few early specs permitted "default" sine tests. Don't be fooled; there is no equivalence between random and sinusoidal tests.

23.10 Shaker ratings. The question "Will our shaker system develop enough force?" is pertinent. Rather than peak A and F, as in Section 12.3 for sinusoidal testing, here we use broad spectrum RMS A and F.

$$\text{RMS g units} = \frac{\text{RMS force, lb}}{\text{total moving weight, lb}} \quad (23.6)$$

$$\text{RMS}\,\frac{m}{s^2} = \frac{\text{RMS force, newtons}}{\text{total moving mass, kg}} \quad (23.7)$$

A shaker's RMS F rating is usually based upon heat generated in the moving coil. Peak sine F = 1.414 × RMS F (based upon 1.414:1 peak: RMS ratio of a sine wave). An additional rating: the "random peak" value, usually equal to 3 × RMS F rating (based on 3σ peaks); see Sec. 22.10.

The purchaser of a system must fully understand how it is rated, as an "underpowered" amplifier may be furnished to lower the selling price. Will the system develop its rated F on all loads? Or only on heavy masses?

Whereas with sine testing there is some concern about exceeding shaker D and V limits, with random testing these limits rarely cause difficulty. We would not attempt seismic testing on an EM shaker of 1 inch D, for example, but would instead use an appropriate stroke EH shaker.

23.11 Splitting a spectrum into two or more sections is sometimes requested, when a shaker's F = MA capability is insufficient. Do not grant permission. We must *simultaneously* excite all resonances in all parts of our spectrum. True, splitting a 20 to 2,000 Hz white spectrum into two equal spectra will require only 0.707 as much F as with the total width. We repeat, this is not an equivalent test.

23.12 Narrow band swept random testing has been used in Europe. Visualize a 30 Hz or 100 Hz band sweeping 50-2,000-50 Hz. As with swept sine testing, resonances will be sequentially excited (rather than simultaneously excited, as with broad band random). However, their responses will be less severe than with swept sine. Fatigue failures will resemble those with broad band random. The major appeal was economy. Equalization per Sec. 25.0 was not needed. Dual purpose AGC regulated either sine or narrow band random.

23.13 Exotic spectra such as sine on random (SoR), random on random (RoR) and gunfire, are difficult to implement with analog equipment per Sec. 25.4. Faster, easier digital techniques are discussed in Sec. 26.5. Gunfire tests are discussed here as an example.

Gunfire vibration tests ensure that aircraft equipment can withstand combined flight random vibration + pulsed over-pressure excitation from machine guns (Gatling guns are particularly troublesome.) Spectral details depend upon gun caliber and firing rate, also the distance from the gun(s).

Consider simulating this excitation via taped signals (Sec. 24.1) from the actual flight environment. Consider also synthesizing this excitation via a broad spectrum (Sec. 23.7), but adding spectral lines. Base the test upon measured vibrations if possible; failing that, -810D offers prediction methods. Interrupt the shaker to simulate short gun bursts.

23.14 Up intensity, down test duration. We often increase test intensity above that normally encountered in transportation or service, to save test time. This is dangerous. Consider carefully all ramifications. Refer to the two-reed demonstration of Sec. 21.7. The likelihood of failure (as with all complex equipment) was extremely dependent on intensity. At very low intensity, the reeds would probably never strike.

Yet we occasionally encounter "rules of thumb" such as: +3 dB reduces test time 10:1. Using such a "rule," one might "compress" 100 hours of flight into 10 test hours. Would +6 dB reduce test time to 1 hour? The "real world" is far from linear. Even the first increase would probably cause highly unrealistic failures. In addition, such "rules" are generally based upon fatigue failures; you are concerned with many additional failure modes.

"Speed up" formulas often assume linearity of structural response to input. That assumption sometimes works, if we can stay within a material's elastic range. But damping is notoriously non-linear. Electronic assemblies (with such non-structural elements as soldered connections and spring-loaded contacts) certainly are not linear.

This is a very complex subject, well beyond the scope of this text. You may wish to read, for example, Lambert's "Criteria for Accelerated Random Vibration Tests with Non-Linear Damping" in Part 3 of SHOCK & VIBRATION BULLETIN 53. And, somewhat simpler, Shinkle's "Automotive Component Vibration: A Practical Approach to Accelerated Vibration Durability Testing," SAE Technical Paper 840501, 1984.

Section 24
Signal Sources for Random Vibration Testing

24.0 **Introduction**
24.1 **Tape playback**
24.1.1 Automotive vehicle tests
24.1.2 Automotive fatigue tests
24.1.3 Not popular in aerospace
24.1.4 Taped random for stress screening
24.2 **Synthesized random**
24.2.1 Frequency domain characteristics
24.2.2 Time domain characteristics

24.0 Introduction. Here we discuss reproducing magnetic taped signals representing vibration, also synthesizing random vibration to meet MIL and other specifications.

24.1 Tape playback. To reproduce taped signals with reasonable fidelity, "boost" certain frequencies while attenuating others, for reasons given in Sec. 25.0. Sec. 25.1 and 26.3 discuss equalization; those considerations apply also to tape playback (or while recording a tape).

During field or transportation investigations, sensors will have been mounted at various "stations" on a vehicle's structure. Vibration signals will have been recorded on board or remotely via telemetry. Taped signals can yield a variety of statistical information, per Sec. 22, that can lead to test specifications and to synthesizing per Sec. 24.2. However, tapes *can* directly supply shaker signals.

24.1.1 Automotive vehicle tests are the principal application. Tapes representing road surfaces become inputs to road simulators: four (or more) EH shakers, each shaking an automobile or truck wheel. (See Sec. 11.4.)

24.1.2 Automotive fatigue tests. Random forcing is replacing the sinusoidal forcing of Sec. 17.1. Sinusoidal vibration road input is never experienced on automotive vehicles. The main reason we used sine forcing for fatigue tests for 100+ years: we could make inexpensive shakers somewhat resembling those of Sec. 10. Road and off-road inputs are generally random. Fatigue test inputs (see Fig. 11.5) generally utilize EH shakers and increasingly apply random forcing. Tape playback is popular but some random signals are synthesized.

24.1.3 Not popular in aerospace. Tape playback is not as popular in the aerospace industry, for several reasons:

1. Assume that six "identical" missiles will be test flown. A tape from flight #1 could be used to test components for missiles #2 - #6. But the tape might be considered too "rough" or too "smooth" — not representative. Or flight #1 might have crashed. Or hardware for missiles #2 - #6 might be built before flight #1.

2. A tape recorded on flight #1 might be available. But designs may have changed.

3. That tape represents vibration at one specific location. Units for a different location demand different "tailoring."

4. Even at the same location, but on flight #1 lacking a certain "black box," vibration differs from that with the unit present. Flight tests could employ "dummy" units, but building a dynamically equal "dummy" is never easy.

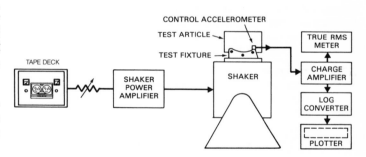

Fig. 24.1 Audio tape cassette scheme for random vibration needed in electronics production stress screening.

24.1.4 Taped random for screening is mentioned in the U.S. Navy's P-9492 (see Fig. 30.3) which suggests a Grumman technique for inexpensive random vibration with an existing sine test system. A $500 audio tape cassette unit records the power amplifier input voltage during a 1g 20-2,000 Hz sine sweep (fixture and load attached). That recording is later used with a noise generator and filters, to synthesize and record a signal for use per Fig. 24.1. (The synthesis facility serves many production facilities.) There are numerous operational drawbacks; for example, changes in the test item to some extent invalidate the tape. Few firms use this approach.

24.2 Synthesized random. In the aircraft/missile field, most random vibration for testing is synthesized. What electrical signal is needed?

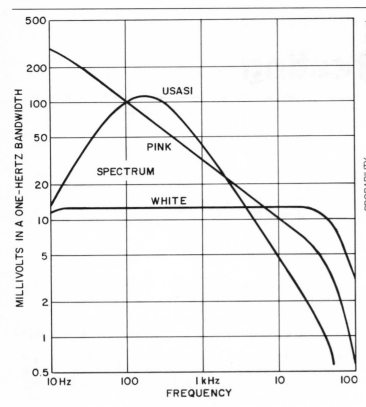

Fig. 24.2 Noise generators offer various spectra. The "white" spectrum, band limited to say 2,000 Hz, is most used for vibration tests. Courtesy GenRad.

24.2.1 Frequency domain. Typical random vibration test specifications were described in Sec. 23. Imagine trying to achieve the spectrum of Fig. 23.2, calling for constant g²/Hz across a wide band. Connect a "random noise generator" (quite common in electronic laboratories) into your PA and shaker system. Limit its spectrum (here 20-2,000 Hz) from

the "white" curve of Fig. 24.2 by a bandpass filter to whatever range is needed for your tests. The essential element of such an analog noise generator is an extremely noisy diode. Digitally generating both true and pseudo random signals is discussed in Sec. 26.3.

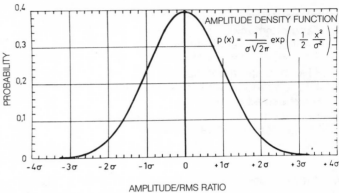

Fig. 24.3 "Gaussian" or "normal" amplitude probability density desired in noise generators for vibration tests.

24.2.2 Time domain. Per Sec. 22.10, the instantaneous amplitudes of random vibration have no regularity. However, they approach the "normal" or Gaussian distribution of Fig. 24.3. So should the signals from your noise generator. As a "quick check," connect it to a long persistence oscilloscope of "Y" sensitivity 1 inch peak to peak per volt RMS. With a 1 volt RMS noise signal, note most brightness within ±0.5 inch of the zero volts line (the beam is in this range 68% of the time.) Intensity will gradually decrease away from the zero volts line. Better, use an amplitude probability density analyzer as suggested by Fig. 22.5 (analog). Most practical, use the computer that digitally controls your test.

Section 25
Analog Equalization for Random Vibration Testing

25.0 Why is equalization required? Imagine that you need an ASD per Fig. 25.1(a) for a 1950s flat or "white" random vibration test. And that a noise generator + an idealized bandpass filter provides your power amplifier with a signal having that spectrum. Will your shaker table ASD be correct? *NO!* Because of resonant effects in shaker and the load (Sec. 14.4), our spectrum resembles Fig. 25.1(b). How can we get (a)?

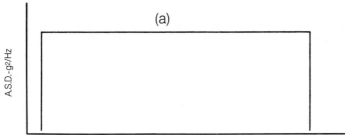

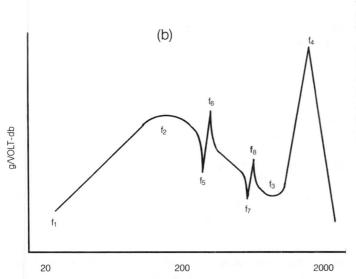

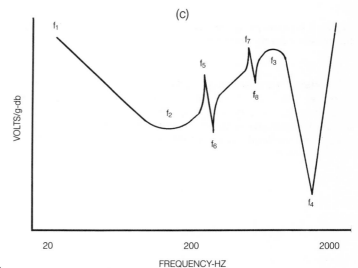

Fig. 25.1 (a) Desired frequency response; (b) System frequency response; see Sec. 14.4. (c) Correction to be achieved through electronic equalization.

For "flat" sinusoidal testing, we varied electrical power to the shaker while varying f_f, increasing power at f_1, f_3, f_5 and f_7, and decreasing power at f_2, f_4, f_6 and f_8, using the technique of Fig. 25.2. That will not work for random, since all frequencies are simultaneously present. We must attenuate our signal at f_2, f_4, f_6 and f_8, and increase it at f_1, f_3, f_5 and f_7, per Fig. 25.1(c), the inverse of (b). (c) + (b) yields (a).

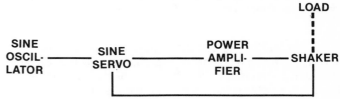

Fig. 25.2 Method discussed in Sec. 14.6, using electronic servo to compensate (one frequency at a time) for shaker + load frequency response.

Shaker and load variations somewhat change Fig. 25.1(b), demanding flexibility in (c). Early analog electronic units are called EQUALIZERS (Fig. 25.3). The process of adjusting these to achieve a desired spectral shape, is EQUALIZATION or EQUALIZING. The spectral shape might include one or more "notches" per Sec. I-4.4 of MIL-STD-810D (Method 514.3), and Sec. 14.8, 23.0 and 30.2.3, possibly to compensate for mobility effects. Today's digital methods are discussed in Sec. 26.

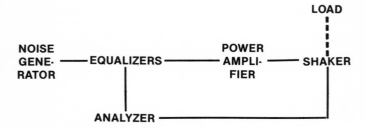

Fig. 25.3 Equalizers compensate all frequencies simultaneously for shaker + load frequency response.

25.1 Band equalization. Consider 1958 manual analog controls per Fig. 25.4, still (1984) being purchased for applications described in Sec. 25.1.1 and 25.1.2, also for tests where test article mass << armature mass + fixture mass. (Single or multiple accelerometers may be used; see Sec.

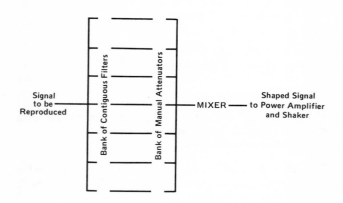

Fig. 25.4 Noise (or tape) signal is divided into frequency windows, typically 10 Hz wide at low f_fs and 100 Hz wide at high f_fs. Then voltage within each window is manually adjusted. Signals are recombined, and (now shaped to compensate for shaker + load) go to power amplifier and shaker.

16.11 and 16.12.) The operator equalizes at a very low level $\approx$ 1g RMS, using a signal per Fig. 24.2 and 25.1(a). Shaker and load vibrate randomly, while he adjusts his attenuators (Fig. 25.4) to obtain the desired spectral shape. Finally, he advances system gain to the specified intensity, and the test commences. He can make minor adjustments to compensate for system nonlinearities. He should make a "proof of equalization" graph (Fig. 25.5) for his test report.

Fig. 25.5 Typical manually-adjusted equalizer console. Resulting spectrum is plotted below. Courtesy Thermotron.

25.1.1 Used for stress screening. Moderate delays in equalizing are tolerable when random vibration is used for stress screening per Sec. 30. This assumes that you equalize, then screen for many hours (or even days) on identical products, in the same axis, without demounting the fixture. Re-equalizing occurs seldom.

25.1.2 Used for reliability demonstration. Moderate delays in equalizing are also tolerable in reliability demonstration testing per Sec. 23. Tests may continue for many days. Re-equalizing occurs seldom.

25.2 Spectrum analysis prior to test. To commence, you must know existing spectral shape, as in Fig. 25.1(b). Then increase or decrease ASD in various spectral windows.

25.2.1 Swept analysis once used narrow-band swept analyzers per Sec. 7.3-7.8 to graph ASD vs. f_f. The operator examined an X-Y plot, changed equalizer settings, then ran another graph. With each sequence requiring about 30 minutes, equalizing was slow, and there was some danger to the test item.

25.2.2 Band analysis, suggested by Sec. 7.9 and here by Fig. 25.6, or by spectral display per Fig. 25.7, more quickly shows the operator the effect of his adjustments. It is con-

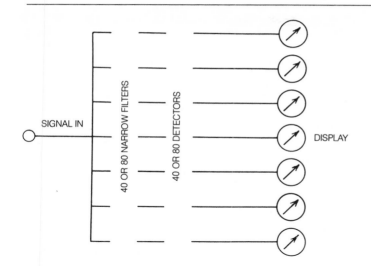

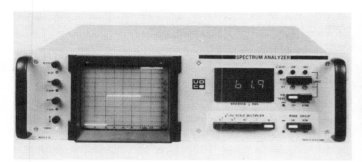

Fig. 25.6 Spectrum analyzer divides signal into frequency windows, then measures g²Hz in each and displays these on meter movements.

Fig. 25.7 Alternative display on CRT permits seeing entire spectrum at a glance. Many laboratories instead use an RTA (see Sec. 7.10.6). Courtesy U-D.

venient if analyzer window frequencies match equalizer window frequencies. Now equalization typically takes 10-20 minutes.

Each "point" on the spectral display, or each meter, represents the magnitude of a narrow-band ac signal, scaled in RMS g units or (more usual) in g²/Hz units. If all NoF (number of filters, typically 36, 50 or 80) display points or meters read the same, the operator is ready for "white" random vibration testing. They will differ for other spectral shapes.

How many NoF must there be? We need to see all peaks and notches so we can compensate for them. Some resonances are extremely sharp. Imagine 2,000 equalizing and 2,000 analyzing filters, each 1 Hz wide, free of frequency "drift." Even if analog technology could do this, imagine the cost. We need narrowest filters at the lower f_fs, where load resonances are sharpest.

25.2.3 Hybrid analyzers, completely independent digital or hybrid RTAs per Sec. 7.10, provide much faster response and far greater detail (NoF perhaps 500) than analog equipment.

25.3 Automatic analog equalization systems appeared in the early 1960s. Manual band equalization demands considerable time, because much damping is necessary in each display, to minimize "jitter" per Fig. 22.8. This damping slows the readout meter or CRT reaction. Time is required (after the operator makes a change) until the readout stabilizes; he often over-corrects or under-corrects. There is also some overlap between adjacent bands; adjusting one equalizer band misadjusts the adjacent bands.

10-20 minutes is too slow for many laboratories. Many people may be involved; delays are expensive. Thus in the early 1960s automatic regulating systems sensed and individually regulated ASD in each band. The human operator adjusts sequentially, but automatic systems adjust simultaneously.

Fig. 25.8 Ling Model ASDE-80, worldwide probably the most popular automatic analog random control console.

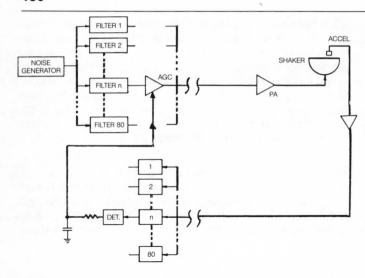

Fig. 25.9 Block diagram applicable to Fig. 25.8. Control console sometimes connects via telephone lines.

Fig. 25.9 suggests 80 control loops. One or more control accelerometers on the shaker table, fixture or test item send a broad-band ac signal to 80 spectrum-dividing filters in the analyzer section, much like Fig. 25.6. Each filter sends a narrow random ac signal to a detector. Each detector generates a dc signal to tell the appropriate AGC unit how much gain is needed, to develop the proper ASD in that frequency window. Then the 80 gain-adjusted signals are mixed into a composite broad-band signal for the PA and shaker.

When the specimen is mounted and ready, the operator slowly advances his master gain control from zero to the proper level for the test, say 10g RMS. At first, all AGC units are at maximum gain (no dc signal from their detectors.) Then a few detectors, covering frequencies f_2, f_4 and f_6 (Fig. 25.1) sense some vibration. Each produces a small dc signal to reduce gain of the corresponding AGC units. As he further advances his master gain control, all detectors produce signal and all AGC units function. The entire process takes less than 10 seconds.

25.3.1 Skill required. While this sounds quite simple, a 1960s operator needed high intelligence and skill. He first read and interpreted his test specification, determining the ASD level for each of his 80 windows. Then he conducted a lengthy "set up" procedure. Loss of such a skilled technician/ operator could incapacitate a laboratory.

Analog consoles changed little from 1961-1971, while random vibration testing was being widely accepted. Difficulties with analog control included filter drift and slowness of response. These problems worsened with narrower (than B = 25 Hz when NoF = 80) frequency windows.

1961 ten-second equalization was much faster than manual sequential equalizing. However, that is too long for random bursts lasting only a few seconds.

25.3.2 Minimum test duration, with automatic analog equalization, is about 30 seconds; typically 80 servo loops must stabilize, the indicators must show the spectrum and the operator must recognize any deviations. One big advantage of digital control: speed. Equivalent digital actions require only about two seconds, with much better spectral detail.

25.4 Exotic spectra such as sine on random (SoR), random on random (RoR) and gunfire, can be implemented with analog equipment. Gunfire tests (see also Sec. 23.13) are here discussed as an example.

Sec. I-4.4 of MIL-STD-810D (Method 519.3) offers two methods. First, the *pulse method.* Develop the required broad spectrum as test item input, using a random console per Sec. 25.3. Then replace the noise generator by a pulse generator (variable pulse width and repetition rate). Adjust signal magnitude to give the desired RMS g. Perform swept filter analysis per Sec. 7.3-7.8 to show that spectral spacing is correct. To simulate 100 firings/sec, set pulse generator pulses at that f_f. The resulting line spectrum should have peaks at 100, 200, 300, etc. up to 2,000 Hz. In some aircraft installations, broadband excitation is negligible, and the gunfire test line spectrum is used alone. If both are present in service, mix the broad + line spectra.

Method 519.3 also offers the *broadband random method.* The controller generates a broad signal having four elevated portions (pedestals) at 100, 200, 300 and 400 Hz. Few if any analog systems can generate such spectra as Fig. 26.11 and 12; they lack dynamic range over narrow f_f widths. Setup time and skill are required for each test. Digital control per Sec. 26.5.3 seems far superior.

Section 26
Computer Control of Random Vibration Testing

Fig. 26.1 Digital random vibration controller. Note CRT terminal. Programs load in about 30 seconds from floppy disk storage. Courtesy H-P.

26.0 Introduction. During the 1970s and 1980s, digital processing techniques increasingly aided random vibration field measurements, also testing and analysis. Prices have dropped. Capabilities have increased. Much analog equipment has been replaced.

Tests with analog equipment went slowly. Lengthy equalization and data analysis, operator errors and equipment malfunction took hours with analog equipment. Minicomputers now generate, control and analyze most random vibration tests. Speed and accuracy are increased, saving manpower. Sec. 26 provides block diagrams, discusses programming a digital control system and briefly explains digital equalization.

Figs. 26.1 and 26.2 show two representative systems, first introduced in the 1970s. Each uses a computer (with associated hardware) to generate and control a random drive signal to excite a shaker and to analyze inputs and responses. System software digitally randomizes and equalizes the drive signal, insuring that all f_f components fall within the specified band and that ASDs are within specified limits.

26.1 Setting up a test requires the operator (after he loads the program) to keyboard test parameters into memory. The computer asks questions; operator answers are underlined in Fig. 26.3.

At line 1, the computer asks for a TEST I.D. to identify test parameters in a program stored earlier. The operator keys that I.D. Within seconds those parameters are again loaded into active memory. At line 2 the computer asks what to title the graphs.

Line 3 BANDWIDTH is the upper frequency f_u desired, here 2,000 Hz. Line 4 asks the desired RESOLUTION or number of windows. This determines the speed and accuracy of

Fig. 26.2 Digital random vibration controller. Note PDP11 computer, RTA and terminal with keyboard and (relatively slow) paper tape storage. Courtesy Scientific-Atlanta.

equalizing a control spectrum (higher values lengthen equalization time but give greater accuracy.) Here we chose 200. (Many analog random controllers provided 80 channels.) The system divides B by resolution and then prints the frequency increment or space between spectral lines.

The required spectrum is defined in lines 5-9, first the initial slope (here +18 dB/octave) up to the first breakpoint (here 20 Hz and .04 g²/Hz; then 50 Hz, .1 g²/Hz; then flat to 1,000 Hz, .1 g²/Hz). The ALARM AND ABORT limits are independently selectable for each breakpoint. Most controllers permit up to 50 breakpoints. Now the operator has finished listing breakpoints and responds to line 9 with a zero. The computer repeated line 9, and asked for the final slope. The operator typed -6 dB/oct (continues to f_u, here 2,000 Hz) and the ALARM and ABORT limits per line 8. The computer calculated and printed the total acceleration level in g RMS. Fig. 26.4 shows the desired spectrum and ALARM lines.

At line 10 the operator limits the f_f range over which the ALARM and ABORT apply (here 20-2,000 Hz). He decides how many spectral lines should activate the ALARM and ABORT limits. If any two reach the ALARM levels specified in lines 5-9, the computer alerts him; if any three reach the ABORT levels, it safely stops the test.

```
ENTER PARAMETERS 1=YES, 0=NO: 1
INPUT 1=KYBD, 2=RT-11: 1

1 TEST ID: DET
2 HEADING: DETECTOR VIBRATION

3 BANDWIDTH: 2000
  BANDWIDTH: 2000.

4 RESOLUTION 100/200/400/600/800: 200
FREQUENCY INCREMENT, HZ=10.00

REFERENCE SPECTRUM:
5 INITIAL SLOPE, DB/OCT: 18
   ALARM LIMITS +DB,-DB:3,-3
   ABORT LIMITS +DB,-DB:6,-6

6 FREQUENCY       HZ.: 20
   LEVEL  GSQR/HZ.: .04
   ALARM LIMITS +DB,-DB:1.5,-1.5
   ABORT LIMITS +DB,-DB:3,-3

7 FREQUENCY       HZ.: 50
   LEVEL  GSQR/HZ.: .1
   ALARM LIMITS +DB,-DB:1.5,-1.5
   ABORT LIMITS +DB,-DB:3,-3

8 FREQUENCY       HZ.: 1000
   LEVEL  GSQR/HZ.: .1
   ALARM LIMITS +DB,-DB:1.5,-1.5
   ABORT LIMITS +DB,-DB:3,-3

9 FREQUENCY       HZ.: 0
9 FINAL SLOPE, DB/OCT: -6
GRMS=12.11

10 ALARM/ABORT RANGE:
   LOW,HIGH FREQ: 20,2000
   LINES TO TRIGGER ALARM: 2
   LINES TO TRIGGER ABORT: 3
11 LOW LEVEL, -DB: -9
12 LEVEL INCREMENT, DB: 3
13 START-UP TIME SEC: 20
14 SHUT-DOWN TIME SEC: 1
15 TEST TIME HRS, MIN, SEC: 0,20,0
16 LEVEL SCHEDULING  1=YES 0=NO: 0
17 AUTOMATIC INCREASE, 1=YES 0=NO: 1
   LOW LEVEL TIME, SEC: 30
18 CONTROL CHANNELS: 1
19 EXTREMAL CONTROL  1=YES, 0=NO: 0
20 AUXILIARY CHANNEL: 2
21 ACCEL SENS MV/G:
   CH 1: 10
   CH 2: 10
22 PRESTORED DRIVE 1=YES, 0=NO: 0
23 DRIVE CLIPPING  1=YES, 0=NO: 0
24 ALARM LEVEL GRMS: 14
   ABORT LEVEL GRMS: 15
25 LOOP-CHECK MAX DRIVE (VOLTS): .3

CORRECTIONS 1=YES, 0=NO: 0
LIST 1=YES, 0=NO: 0
SAVE 1=YES, 0=NO: 1

DEVICE: DX
```

Fig. 26.3 Dialogue between computer and operator. Operator responses are underlined.

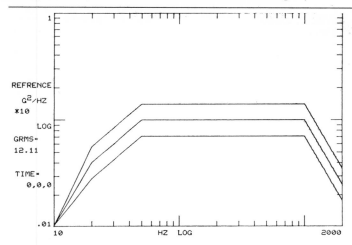

Fig. 26.4 Example of a standard spectrum, as it would be printed out before a test.

At lines 11 and 12 the operator indicates overall g RMS pretest vibration levels, here starting at -9 dB (below full intensity) and increasing in +3 dB steps.

START-UP TIME (line 13) is the interval (in seconds) between the level of line 11 and full intensity. SHUT-DOWN TIME (line 14) is the interval over which vibration drops -30 dB at any ABORT, operator or "test complete" shutdown. The TEST TIME (line 15) applies to the full intensity interval.

LEVEL SCHEDULING (line 16) permits changes in test level and duration, keeping the same spectrum shape. If the operator answers yes (keyboards a 1) the computer asks his LEVEL SCHEDULING wishes.

A yes (1) answer at line 17 calls for automatic stepping through pretest levels per lines 11 and 12. A no (0) response requires manual increases. At line 18 he selects up to 16 input control channels for averaging or limiting.

A no (0) response at line 19 selects the usual single channel PSD control method. (1) calls for extremal control for comparing the response spectra from all control channels per line 18; the highest response over each f_f window now controls ASD in that window.

At line 20 (AUXILIARY CHANNEL) the operator selects a channel for observation during testing. (Any one of 16 input channels can be reset during the test.) He can print out any spectrum.

At line 21 he states control and response accelerometer sensitivities in mv per g for each channel number. Line 22 permits full test intensity after a "stop," using the most recent drive signal from the previous test (stored in memory) as when repeating tests with identical payloads.

For DRIVE CLIPPING, he answers yes (1) at line 23; then he must state the clipping level. (See Sec. 27.4.)

He inserts overall ALARM and ABORT GRMS LEVELS at line 24. (Don't confuse with the ALARM and ABORT LIMITS earlier selected for each breakpoint.) Line 24 applies to the entire spectrum. If an ALARM LEVEL is reached, the system notifies the operator. If an ABORT LEVEL is reached, the

system stops the test. At line 25 he selects a safe drive signal magnitude (if no feedback signal is received at startup) so the system won't "run away".

After he loads test parameters into memory, he can list all or part of the program on his line printer or CRT terminal by answering 1 at LIST. He can correct (edit) individual test parameters without changing others. "SAVE" refers to paper tape or disk storage.

Finally, he calibrates the controller by simply pushing a button. The computer generates a "Standard Spectrum" with CRT display resembling Fig. 26.4. A hard copy is available within seconds. The system is now ready.

26.2 Running the test is easy. Connect all external instrumentation, energize the system, press the start button, sit back and watch. The controller applies a small drive signal per line 25 for a "loop check". If there is no feedback signal, the computer aborts and flashes or prints an appropriate fault code such as "OPEN CONTROL LOOP." Some systems warn if the control signal is too noisy.

If all is well, the controller increases overall g RMS, stopping to equalize the spectrum at each pretest level. You could assume manual control. The computer continually monitors the spectrum for any anomalies or "out of spec" conditions.

The timer starts when full test intensity is reached. The controller continues to equalize the spectrum until time expires, you stop the test, or an abort occurs.

At an abort, the controller displays a fault code or prints the reason. Most common: GRMS ABORTS (the overall g RMS level has exceeded limits); LINE ABORTS (one or more spectral windows have exceeded limits) and LOSS OF CONTROL SIGNAL. Check for loose or broken accelerometer leads; loose or intermittent accelerometer; ground loops; loose or cracked fixture or test item or a bad signal conditioner. A poor fixture may prevent equalization (see Sec. 19).

Most controllers also provide for external safety aborts (e.g. shaker overtravel or overheat or PA overheat or test observer safety switches).

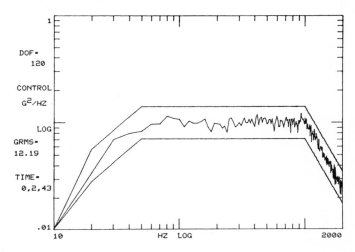

Fig. 26.5 Example of a control test spectrum, as it would be printed out during or after a test.

Some controllers announce equalization accuracy during a test, calling for more PA gain if needed. If ALARM limits are exceeded, the computer advises which window, + or -, and how much.

During a test, you can continually observe ASD from the control accelerometer, along with overall g RMS and elapsed time. Before, during and after a test, you can store and/or generate a hard copy, fully annotated, of any control or response spectrum, as in Fig. 26.5. To obtain a response during a test, put the controller in "OPEN LOOP," not equalizing the spectrum but maintaining the last drive signal. Disconnect the control accelerometer signal and feed in a response accelerometer signal.

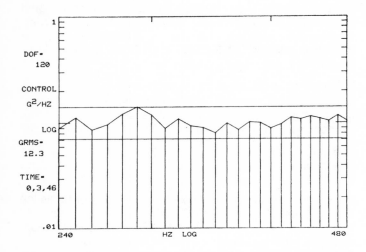

Fig. 26.6 Hard copy of a spectrum "zoom" expanded to better show a particular portion. Optional frequency bars delineate windows.

Many systems have a CURSOR and EXPAND feature. With this, you can select and view any f_f band on your CRT and get a hard copy per Fig. 26.6.

Digital systems can display and/or reproduce other functions before, during and after a test:

1. Transmissibility plots such as response/input.
2. Out-of-band data beyond f_u, say 2-10 kHz.
3. The drive signal in the time or frequency domain.
4. The error spectrum.

Under a MODIFY function, you can change the spectrum *during* your test. Your line printer or terminal will ask for the band to be modified and the level increment in dB. Between the chosen limits, you can step the reference level up or down, and shift the frequency band left or right. The new parameters are promptly equalized.

26.3 How the system works. Let's examine the major components and the equalization process. Fig. 26.7 shows a basic system's simplified block diagram. Major functions: generate and control a random shaker drive signal and analyze the resulting responses. Major activities:

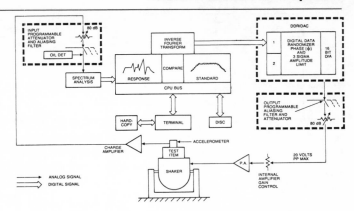

Fig. 26.7 Simplified block diagram of a digital random system. Note that some signals are analog, some digital. Courtesy Scientific-Atlanta.

1. *The minicomputer*, the system's heart, contains the appropriate peripheral hardware and memory (usually 28K). As operator, you speak to it through a keyboard. It speaks to you via displays and/or printouts, also with system hardware through special Input/Output (I/O) boards. It performs all calculations and stores information in memory.

2. *The system control unit* contains logic circuitry, input and output amplifiers, and the output low pass filter.

3. *The spectrum analyzer* accepts the control accelerometer signal and periodically converts that time-domain input to a frequency spectrum.

4. *The line printer* or *CRT terminal* (keyboard + cathode ray tube + hard copy device) links the operator and computer via a tape reader, floppy disk and/or keyboard.

Your system (1) generates your shaker drive signal, and (2) eliminates the effects of your shaker transfer function (as in Fig. 14.13).

When you enter g^2/Hz values (your *reference spectrum*) into memory, the computer internally generates its *standard spectrum*. The first low-level attempt at developing a drive spectrum (input to PA) reproduces the standard spectrum's shape but at reduced intensity. It ignores (temporarily) the shaker/load transfer function, which will eventually be equalized. Then its values will be inverse Fourier transformed to the time domain.

At this point the time-domain drive signal is pseudo-random, repetitive, for a fast estimate of the system's transfer function. For true random vibration testing, it must be randomized (in the DDR/DAC — Digital Data Randomizer/Digital to Analog Converter), which produces a stationary Gaussian continuous spectrum drive signal.

Most controllers allow selection of pseudo-random vs. true random tests. With true random, your time history is Gaussian (see Fig. 22.4) with a peak/RMS ratio up to 4.5; or you can clip at $\pm3\sigma$. With pseudo-random, you can equalize faster and with greater accuracy. Possible objections to pseudo-random operation: 1) the amplitude histogram is not quite

Gaussian and 2) "it sounds different." Some systems allow change from pseudo-random to true random during pre-test level equalization. This helps with massive shaker loads having many high-Q resonances. Also on very brief (little equalizing time) tests; here you specify pseudo-random during pre-test equalization and true random at test levels. (Here we discuss true random testing.)

The generated signal now goes to the output attenuator, filter, PA and shaker. Shaker motion causes control accelerometer/charge amplifier response at the System Control Unit. Here the computer selects proper attenuation or gain to optimize dynamic range.

Now the shaker motion signal goes to the spectrum analyzer (which FFTs it into the frequency domain) and thence to memory, where a predetermined number of spectra (an ensemble) is accumulated and averaged, producing the *table spectrum*. (It differs from the drive spectrum due to the shaker system transfer function.) The computer compares the table spectrum with the initial drive spectrum and generates a new drive spectrum. This is IFTd (or inverse FFTd) to create a new drive signal.

Some systems provide optional drive clipping (see line 33 of Fig. 26.3). Clipping extends your spectrum, possibly beyond f_u, your upper test frequency. Per Sec. 27.4, this argues against clipping.

Each equalization attempt requires an interval called the system "loop time," up to 9 seconds, per Table 26.1. During that interval nasty things *can* happen; for instance, the control accelerometer *can* drop off. The system *can* lose control. These possibilities promote protection per Sec. 27. (Salesmen might not mention this.) Loop time is the summation of times required to:

1. Fourier transform for PSD measurement,
2. Compare result with the reference spectrum,
3. Generate a new drive spectrum and
4. IFT to a new time domain drive signal.

The equalization process continues throughout pretest and test levels.

Loop time is an important parameter. Checks for alarms, levels and aborts, also spectral corrections for equalization, only occur once per loop. Fast loop time is often desired for speedy updating and frequent protection. Since loop time includes control ASD measurement, high equalizing accuracy (large number of statistical DoF) may require extending loop time. Higher DoF requires additional averaging (more independent measurements) per loop. Control accuracy is traded against loop response time. One tradeoff control algorithm is based on N, the number of averages per loop and on K, an exponential weighting variable. N increases accuracy while K determines fast system response to dynamic changes. Both affect the DoF but only N affects loop time. One digital control supplier gives statistical accuracy of the control spectrum (in DoF) by the formula:

$$2N(2^{K+1} - 1). \qquad (26.1)$$

Some systems have software pre-programmed with N = 4 and K = 3, yielding a control spectrum with 120 DoF and

about ±1 dB accuracy, acceptable for most tests. However, you can change DoF via changing N or K. Some systems ask this during setup. Other systems provide a switch. Others ask for DoF required; after you respond, the computer calculates the necessary equalization parameters and prints out PSD tolerances and abort levels. Table 26.1 shows loop times for a typical digital random system (different computers on two dates). Loop time varies with system B, resolution and N.

Table 26.1
TYPICAL LOOP RESPONSE TIMES
Courtesy GenRad

BANDWIDTH (HERTZ)	RESOLUTION (# OF LINES)	FRAMES PER LOOP	LOOP TIME (SECONDS) 11/04	11/34
2000	100	N = 1	0.8	0.7
2000	200	N = 1	1.6	1.3
2000	200	N = 2	1.7	1.4
2000	200	N = 4	1.9	1.6
2000	200	N = 8	2.3	2.0
2000	200	N = 16	3.2	2.8
3000	200	N = 4	1.8	1.4
5000	200	N = 4	1.8	1.3
8000	200	N = 4	2.0	1.5
2000	100	N = 4	1.0	0.8
2000	400	N = 4	3.8	3.0
2000	800	N = 4	9.0	7.2

Another digital control manufacturer uses the same N and K symbols but with nearly opposite meaning. Note that N here differs from N used elsewhere in this text (symbol for force in newtons) and from the N in Sec. 7.10.6. And that K here differs from K used elsewhere in this text (symbol for spring stiffness).

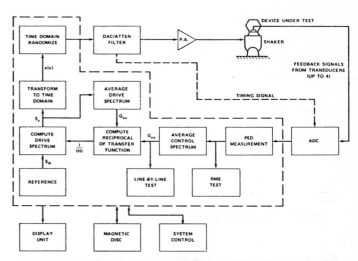

Fig. 26.8 Block diagram of a digital random system featuring FFT processing. Courtesy H-P.

All suppliers' equalization processes are similar, trending today to all-digital processors and the FFT algorithm, away from hybrid RTAs. Fig. 26.8 and 26.9 show other random vibration controllers. Several channels can be processed for averaging or limiting. Here the ADC precedes ASD measurement. FFT results are compared to the reference spectrum. A new drive spectrum is computed and IFTd to the time domain.

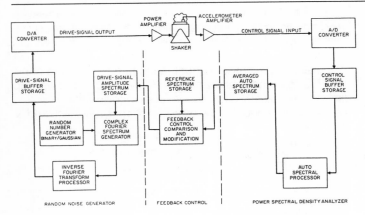

Fig. 26.9 Block diagram of a digital random system featuring FFT processing. Courtesy GenRad.

Although hybrids, per Sec. 7.10, offer high speed digital memory, economy and reliability, an FFT permits other tests and analyses. As later discussed, additional software permits other shaker tests such as shock, shock spectrum and sine + random. FFT systems can also utilize additional accelerometers for tests requiring averaging or limiting. Also, FFT preserves data phase as well as magnitude, permitting auto-correlation, cross-correlation, transfer functions, phase measurements, Co-Quad and coherence analysis.

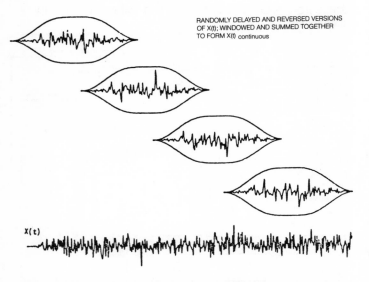

RANDOMLY DELAYED AND REVERSED VERSIONS OF X(t); WINDOWED AND SUMMED TOGETHER TO FORM X(t) continuous

Fig. 26.10 Time domain randomization. Successive delayed and occasionally reversed versions of the waveform produce random signals; they are multiplied by a time window and overlapped to form a continuous signal. Courtesy H-P.

During the test, tape record various inputs and responses. Then after the test, load special analysis software into computer memory. Bring in any taped signal (to the input connection where earlier the control accelerometer signal arrived). Now you can perform auto-correlation, etc. At least occasionally check amplitude probability density (APD), similar to Fig. 22.4.

26.4 Options aplenty are offered, such as simultaneous remote control of up to 8 shakers with one computer. An on-line RMS PROTECTION option monitors each shaker (independent of control per Sec. 26.3). Upper and lower RMS abort limits (programmed during input dialog) are needed; while the computer is regulating one shaker, the others have on-line RMS shutdown protection. Another option available is a 16 channel signal selector (MUX) for peak response limiting, averaging or data acquisition.

26.5 Exotic spectra. Some manufacturers offer "advanced" spectra for MIL-STD-810D and other testing. Random on random, sine on random and gunfire tests have mainly been developed by users. Their work may be available through your system's manufacturer.

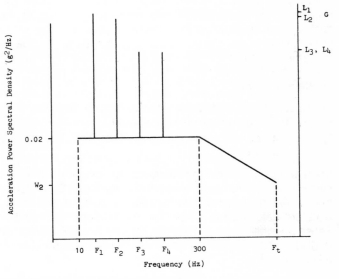

Fig. 26.11 Sine on random spectrum.

26.5.1 Sine on random (SoR) tests require single or several fixed or swept sine vibrations combined with broadband random, per Fig. 26.11. Load special SINE ON RANDOM software and key the necessary test parameters.

Performing such a test generally follows Sec. 26.2. The operator keys sine sweep rate, frequency limits and "g" levels. The test starts with a system loop check.

The system inverse FFTs each drive spectrum. The two time domain signals are summed for the DAC to produce an analog shaker drive signal. Shaker response determines the system transfer function, used to estimate both control spectra. Both drive signals and both control estimates are independent.

Some systems feature special hardware for sine + random tests. A digital sine controller interconnects with a digital random controller; independent drive signals are summed in the system mixer.

26.5.2 Random on random (RoR) tests require one or more narrowband random signals sweeping on a broadband spectrum. You load a special program and key test parameters. GenRad offers spectra per Fig. 26.12.

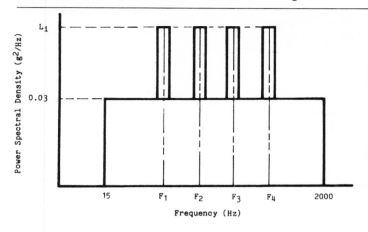

Fig. 26.12 Random on random spectrum.

RoR control resembles Sec. 26.3 with several spectra. You can run a broadband spectrum alone or combine narrowband and broadband spectra. The combinations can be defined as either the maximum of the broad and narrowband spectra (whichever is greater at a given f_f) or as the sum of the two spectra. For the narrowband spectrum, key sweep rate, f_l and f_u.

Alternately, a digital sine controller (can also handle narrow random) and a digital random controller can be interconnected to perform RoR tests. This, as with SoR, requires the purchase of several options. The narrowband and wideband random drive signals are summed in the system mixer.

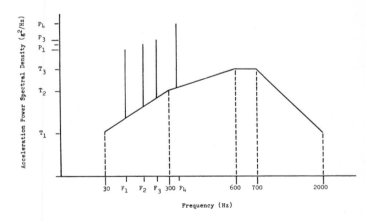

Fig. 26.13 Generalized gunfire vibration spectrum shape.

26.5.3 Gunfire vibration tests (see Sec. 23.13), are difficult with analog techniques per Sec. 25.4. Digital techniques aimed at the spectrum of Fig. 26.13 are still rather crude, but they offer several advantages over analog control: (1) greater number of filters for better spectral control, also (2) programs can be saved, speeding setup of future tests. At press time, dedicated software packages have not been released. One supplier offers an interim gunfire procedure, utilizing transient vibration control software per Sec. 33.4.1. The operator keys in his desired spectrum; the computer IFTs this to the time domain for disk storage. Preparing for a test, he keys

pulses/sec and total number of pulses (as in a gunfire burst). See Cies, "Gunfire Vibration Simulation on a Digital Vibration Control System," SHOCK & VIBRATION BULLETIN 52.

Fig. 26.14 Another modern digital random system. Note CRT terminal and floppy disk storage. Courtesy GenRad.

Fig. 26.15 Another modern digital random system. Note CRT terminal and display, floppy disk storage, printer and optional controls. Courtesy Scientific-Atlanta.

26.6 Recent developments. Fig. 26.14, 26.15 and 26.16 show some modern vibration controllers. All can control a wide variety of tests: random, sine and shock. They can perform a wide variety of analysis functions.

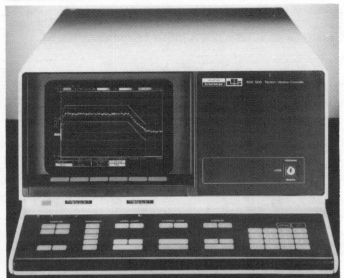

Fig. 26.16 Compact digital vibration controller (random only) uses the Winograd FFT algorithm. Up to 99 "break points" can be selected among up to 8 different test profiles. Courtesy LDS and Solartron.

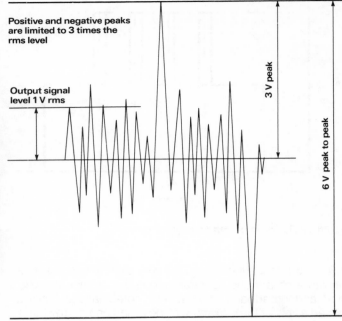

Positive and negative peaks are limited to 3 times the rms level

Output signal level 1 V rms

3 V peak

6 V peak to peak

Fig. 26.17 Example of digital random system (Fig. 26.16) output signal. Note crest factor, here 3 (widely accepted), but variable from 1.4 (useful when quality is not specified) to 8 (useful to minimize out-of-band signals, per Sec. 27.4). Courtesy LDS and Solartron.

Most of today's systems store data in fixed-length blocks on soft-sectored, 8-inch floppy disks. Each can store up to 512 kilobytes of data and transfer at 62 k-bytes/sec. Replace earlier paper tape readers.

26.7 Justifying digital random control. Most lab managers agree with the benefits of digital random control, but may not be totally "sold." Some bemoan lack of specially trained personnel to troubleshoot, maintain and calibrate. However, today system suppliers:

1. Offer system service contracts, including regular maintenance and calibration checks, plus emergency service.

2. Furnish diagnostic and calibration software. A diagnosis is displayed and/or printed to tell the operator which (if any) section is malfunctioning. Calibration software tells how to adjust the system. Instructions, readings and "out of spec" conditions are displayed.

3. Calibrate, maintain and repair digital vibration systems. Many problems can be diagnosed and repaired by phone.

Controller size is dropping. Compact controls move easily from shaker to shaker for various tests or analysis functions.

Section 27
Protecting the Shaker and Specimen

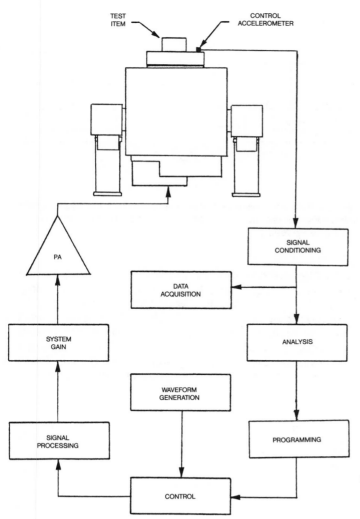

Fig. 27.1 General block diagram — typical electromagnetic shaker system.

27.0 Introduction. Here we deal with clippers, protectors and techniques that protect the shaker and test specimen. Many engineers overly trust their modern solid state instrumentation, PAs and digital controls, doubting need for additional protection. The malfunction probability is small, but many errors and malfunctions *can* occur in a complex system per Fig. 27.1. The overall costs (test invalidation, test specimen damage, shaker damage, delays, retest costs, etc.) can be high. Possible problem sources: power failure, instrumentation malfunction, PA malfunction, calibration errors and operator errors.

27.1 Protecting the shaker. The PA can generate large mechanical forces in the shaker armature, driving it hard against mechanical stops. This may damage a valuable test item *and* the shaker. Problems such as flashover occurred within early tube-type PAs. Even with solid-state PAs, line voltage surges and transients can occur. Limiters ahead of and within the PA cannot protect against all faults.

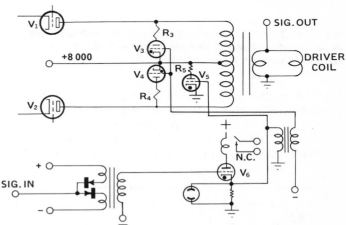

Fig. 27.2 Concept of early thyratron and ignitron shaker protector for high voltage electron tube amplifiers.

27.2 Dangers to abrupt power amplifier shutdown. The use of solid state PAs (see Sec. 13.3) has reduced malfunction probability. However, operator error can surge output voltage and/or current. A shaker overtravel switch can trip. Various factors can abruptly stop tests. Field changes (as when a contactor opens in the PA) may reach the PA input (and thus reach the shaker) before the shutdown sequence ends.

Fig. 27.3 Shaker and specimen protector for sine or random testing. Courtesy U-D.

27.3 Protecting the specimen is the goal of various accessory protectors, also safety features of digital controllers. They function when a previously-set intensity is exceeded or the operator errs or the system malfunctions. The unit of Fig. 27.3 monitors various PA and shaker parameters. Operator adjustable sensor level #1 monitors PA currents, voltages and temperature, also shaker overtravel. Above a certain level the device compresses the signal, but the PA remains "on." Fixed sensor level #2 triggers on excessive PA output current and voltage, several temperatures and coolant flows, also shaker overtravel, initiating a shutdown sequence which: (1) compresses the vibration signal, (2) disconnects the output stage from the preamplifier and (3) cuts main power. Thus shaker transients are not generated. System capability can be restricted for low power tests.

Fig. 27.4 Vibration monitor-protector used for both sine and random testing. It shuts down test if predetermined intensity (D or A) is reached. Courtesy U-D.

Fig. 27.4 and 27.5 show and Fig. 27.6 block diagrams control loop protection. The units trip if there is no acceleration signal from the shaker. The operator presets them to trip on reaching an RMS or a peak level. Then they smoothly compress the signal and indicate the cause: high or low

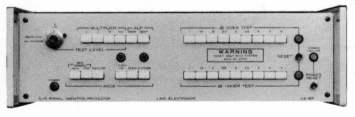

Fig. 27.5 Another monitor-protector for sine or random testing. Courtesy Ling.

level. They can also be interconnected with sine servo systems. Thus no signal will reach the shaker when the protector is reset. On multi-level sine tests they provide for program step changes.

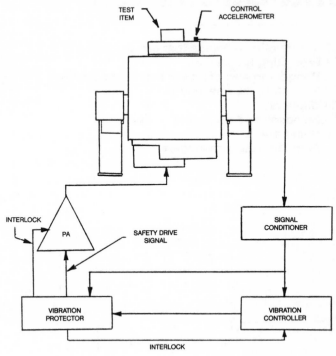

Fig. 27.6 Block diagram of vibration protector used to safely shut down the the vibration controller/servo and deenergize the power amplifier.

27.4 Clippers operate upon low-level signals, ahead of the PA. Use is dictated more by test standards than by safety. They affect and distort, but never terminate the test. They typically remove $\ddot{x}$ peaks $> 3\sigma$. Peaks up to 5σ or 6σ occur in noise generator signals. Many test standards require $\ddot{x}$ peaks to at least 3σ. If you fear to exceed 3σ, use a "clipper" to block the few peaks greater than $\pm 3\sigma$ from reaching your PA and shaker, yet pass smaller $\ddot{x}$. With a marginal PA, clipping may not be needed.

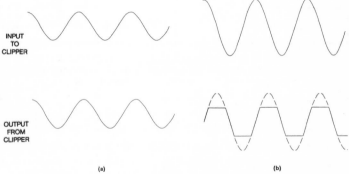

Fig. 27.7 Effects of clipping a sinusoidal waveform.

Fig. 27.7 shows analog clipping changing a sine wave (difficult to show on a random signal). Small signals pass unchanged. As with clipping in some square-wave generators (to convert sine waves into square waves) clipping introduces much distortion. Some digital systems offer a clipping option; see line 23, Fig. 26.3.

Be careful when you opt for clipping; it generates a new spectrum. Examine your spectrum to say 10 kHz when testing to say 2 kHz f_u. New f_fs lie above f_u, outside the control range; severe clipping changes drive signal statistical properties and may prevent proper equalization. Clipping at ±2.5 or 3σ is usually OK.

27.5 Selection and averaging - digital vibration controllers (see Sec. 15 and 26) can select between and can average up to 16 channels for both sine and random testing. For sine testing the program averages the level(s) keyboarded by the operator. (Selection and averaging via analog techniques are discussed in Sec. 16.11.)

Random tests under "extremal" control compare the signals from several accelerometers. Whichever has the highest ASD in each f_f window is used for control in that window. Such tests are extremely gentle and may be meaningless. For random tests under "average" control, the ASD at each accelerometer is calculated over each f_f window, averaged, and used for control in that window.

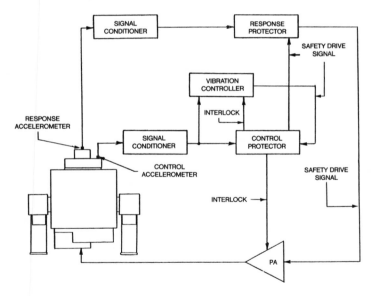

Fig. 27.8 Block diagram showing drive signal routed through two protectors in control circuit. The control and response signals are being monitored during either sine or random tests.

Although clippers, selectors and averagers are primarily intended for control, they can also protect — limit specimen responses. Simply replace a control signal with a response signal. Better, use two protectors per Fig. 27.8 in series with the safety controlled drive signal. You *can* overtest and undertest protect at the control point and overtest protect at the response point. See also Sec. 26.4.

27.6 Summary. With today's digital vibration controllers, some question the need for additional protection. They trust their controller's safety features. Some ignore "loop time" (Sec. 26.3), long intervals without ASD control in a particular f_f window (a damaging resonance might build up.) Consider "backup" in case your controller malfunctions. Note that digital control attempts to compensate for faults, possibly resulting in severe overtest or undertest.

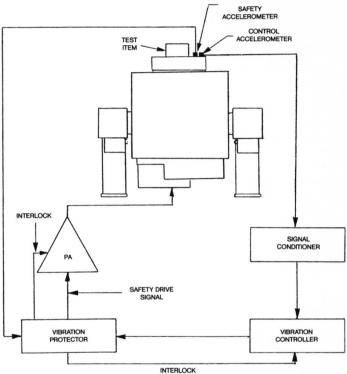

Fig. 27.9 Block diagram showing a completely independent monitor/protect safety channel. The vibration controller/servo and the vibration protector should closely agree.

Mount a separate "safety" accelerometer near your control accelerometer. Connect it to your vibration protector per Fig. 27.9, providing a completely independent monitor and protection data channel. If they differ, seek the cause.

27.7 Alert laboratory managers recognize that accidents can happen during setup and operation. Even careful operators can err. We recommend that you use these simple operating techniques and procedures:

1. Regularly maintain (preventive) your system, especially the PA and shaker. Torque all terminal strip screws. Check all electrical wires for frayed ends, for continuity and for leakage. Regularly change filters. Check cooling systems. Operate your system per the instruction manuals and Sec. 13.10. Log armature current, cross-talk, etc.

2. Regularly calibrate all accelerometers, signal conditioners, limiting and protecting devices, vibration controllers, meters, etc.

3. Use separate monitoring and analysis channels (completely independent hardware from accelerometer to analysis) to verify test results. Monitor time waveforms on a 'scope to detect any abnormal conditions.

4. Supervisors or test engineers should write clear and detailed operating procedures. Cover everything from mounting accelerometer(s) to programming and starting a test. (To lessen that writing task, refer often to specific sections in equipment manuals and in this text.) Pay special attention to startup procedures; most problems occur when equipment is first energized. Identify all equipment items, all knobs, switches and meters. Place guards on controls that should not be touched.

5. Occasionally surprise your test operators with a drill or a simulated equipment fault. Check their ability to understand and to follow your instructions.

6. Rehearse each test profile, especially after significant system changes. Be realistic: use a dynamically similar load, and system in standard test configuration.

7. Evaluate your control instrumentation before a test at minimally one f_f against an optical wedge (Sec. 3.1.3) near the control accelerometer. 44 Hz and 10g are easy to remember; the wedge should read 0.1 inch D.

We suggest you read Barlett, "Protection of the Test Article in the Vibration Laboratory," 1978 IES PROCEEDINGS.

Section 28
Earth Motion
Measurements
and Seismic Testing

28.0 Introduction. This section presents material of possible interest to environmental test personnel, this book's primary audience. Measuring the effects of man-made disturbances such as blasting (Sec. 28.1), geophysical exploration (Sec. 28.2) and earthquakes (Sec. 28.3) somewhat resembles measuring vibration (Sec. 3-6) and shock (Sec. 32). Sec. 28.4 examines seismic testing (great concern to nuclear power plants), resembling random vibration tests.

28.1 Blast measurements. Over 11 million pounds of industrial explosives are used daily in the USA, for highway construction, rail cuts, quarries, tunnels, etc. Waste energy (earth vibration and airborne noise) arouses fear and sometimes damages glass and plaster. Heavy construction and industrial machinery, also railways and trucks to a lesser extent, can cause similar damage. Entire buildings are sometimes isolated (see Sec. 1.1.5).

Explosives users try to limit earth motion, to avoid violating governmental codes and to avoid citizen complaints. But no single number in a code (such as V) adequately quantizes damage potential. Criteria should include f_ns of nearby structures and the f_f content of ground motion (affected by soil properties).

You must measure to show compliance with criteria. Velocity sensors (per Sec. 4) are popular, and sometimes used in all three axes, in soil and on nearby buildings. Hundreds or thousands may be used on a large test. $\dot{x}$ and x readouts (with integration per Sec. 4.4 and 4.6) are common. Signals and timing marks are usually field recorded on magnetic tape cassettes, sometimes unattended. In the lab, these tapes are analyzed and various readouts are plotted for comparison with criteria. Other readouts may include $\ddot{x}$, spectra, vector sum, etc.

28.2 Geophysical exploration mainly depends on man-made earth motions. Exploring for oil and gas deposits is a well known application. Architects, civil engineers, road and

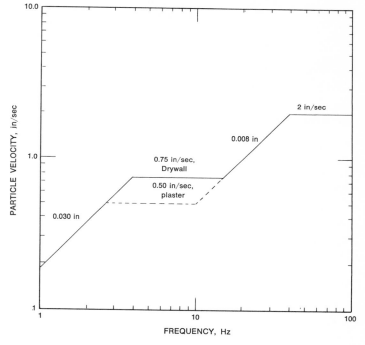

Fig. 28.1 Safe soil motion levels for blasting near residential areas, per U.S. Bureau of Mines Report RI-8507 (Nov. 1980).

bridge designers and contractors, mine and quarry operators also use vibration responses.

Stimulus methods on land include dynamite, tamping (as in Fig. 28.2) and EH shakers (Sec. 11) called servo hydraulic vibrators. The latter are expensive but they are highly versatile, as their spectral content is easily adjusted; sine sweeps are also used. The cylinder normally attaches to a motor truck or lorry, driving the piston downward against a base plate on the soil. At sea, stimulus methods include explosives and charges of compressed air.

Fig. 28.2 Vacuum and gravity pull 80 lb. piston against baseplate resting on earth. Courtesy EG&G Geometrics.

Responses (usually $\dot{x}$ is sensed) are measured and analyzed by much the same techniques discussed in Sec. 2.7, 4.4, 4.5, 5.6, 5.13 and 7. Oscillographs can record $\ddot{x}$ vs. time (see Sec. 7.12) or $\dot{x}$ or x vs. time. Time domain studies are most used.

To mount a sensor to measure seismic vibrations on a building or traffic vibrations on a bridge, use an adequate

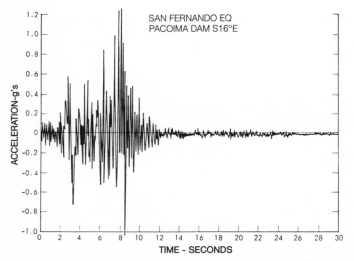

Fig. 28.3 Accelerogram from Los Angeles earthquake February 9, 1971. Large $\ddot{x}$ over time develops large x. After Bernreuter, Norris and Tokarz.

bolt or structural member. But connecting to soil is difficult. See Nolle, "High Frequency Ground Vibration Measurement," SHOCK & VIBRATION BULLETIN 48, Part 4. Thermally insulate the sensor and protect it from wind.

28.3 Measuring seismic events or earthquakes. Most earth motion occurs at low frequencies (typically 0.5 to 50 Hz). Absolute displacement sensors (Sec. 3) are not practical because they require a fixed reference platform. Velocity sensors (Sec. 4) are useful. Accelerometers, particularly servo units (Sec. 5.6), are sometimes used. At maximum sensitivity the micro gs of distant events give useful signals.

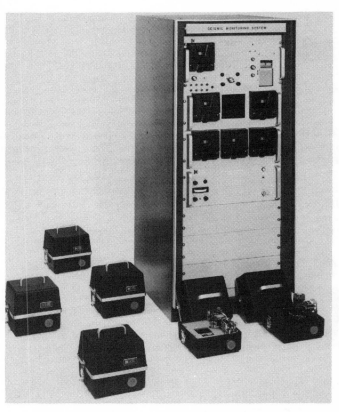

Fig. 28.4 Earthquake monitoring system. Rather than gather oscillograms (per text), centralized unit gathers data for spectrum analyzer. "On line" data gathering + analysis + decision criteria saves even more vital time. Courtesy Kinemetrics.

Immediately after a nearby earthquake, nuclear plant operators need to know severity of effects. Their question: "Need we shut down?" Instrument packages containing orthogonal servo accelerometers and an oscillograph are located close to vital equipment. Following an event, oscillograph records must be gathered, processed and examined. If any $\ddot{x}$ limits have been exceeded, the plant must "cool it." Such a simple criterion must be very conservative. It is misleading, since damage does not relate solely to $\ddot{x}$ peaks. Better to FFT signals to obtain spectra for comparison with a more liberally set limit (based on damage potential). Only when damage is likely should a plant be stopped.

28.3.1 No equivalence to Richter scaling. Peak $\ddot{x}$ of about 0.2g are common close to earthquake epicenters. But $\ddot{x}$ has no relation to the Richter scale, which deals with energy released during an earthquake. An event of Richter magnitude 6.0 might exceed 0.2g motion nearby, but only a micro g across the earth.

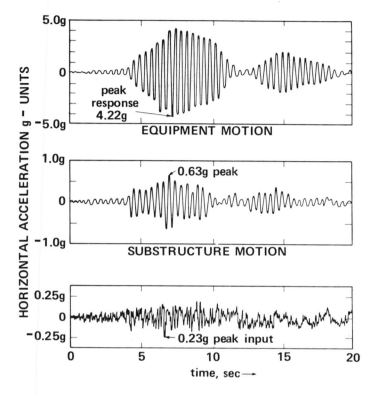

Fig. 28.5 The lower trace represents ground motion during the Richter 6.5 Los Angeles area earthquake of February 9, 1971. After Fischer.

The lower trace of Fig. 28.5 was recorded several miles from the epicenter. The upper traces indicate how a building structure and a mounted equipment might have responded to that input, had a designer carelessly "stacked" resonances, with $f_{equipment} = f_{building}$.

28.3.2 Motions are random. Note the similarity between the input/responses of Fig. 28.5 and those studied earlier in Sec. 21 and 22. Instantaneous magnitudes vary randomly, unpredictable in detail; however, their statistical properties are close to normal or Gaussian.

28.4 Seismic testing via shakers. Even primitive people fear earthquakes. Civilized peoples worry about damage to dams, houses, bridges and buildings, especially nuclear power plants. Stringent seismic qualification requirements apply to the latter, especially structures, piping and equipment essential to reactor shutdown. These must be proven capable of functioning properly during and after imagined Operating Basis Events (OBE) and Design Basis Events (DBE). Standards often call for simulating a DBE following several OBEs. The goal is to simultaneously excite all resonances of the test article, as would an earthquake.

Tests on small equipments may require only a single EH (see Sec. 11) shaker, but those on large equipments may require two or more independently controlled EH shakers driving a platform with test load attached. The platform should be vibrated in the vertical and at least one horizontal axis.

Small EM shakers per Sec. 12 are suitable for low force vibration ("point" input to equipments and structures) to determine f_ns and mode shapes. But for seismic tests, most labs utilize EH shakers per Sec. 11. EH shaker high F, long D and f_f range (0-50 Hz is adequate) are quite suitable. Some tests have used explosives, but they are weak at low f_fs.

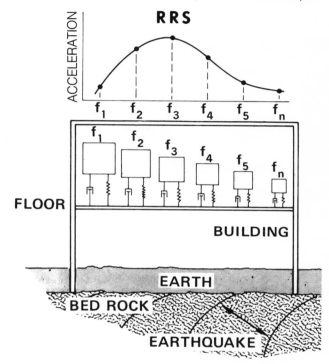

Fig. 28.6 Required Response Spectrum (RRS). After Fischer.

28.4.1 Seismic test requirements may use a tape recording of an actual earthquake, but are more commonly presented as a *shock response spectrum* (SRS - see Sec. 32.6). Interestingly, Biot's 1931 SRS idea was aimed at earthquake

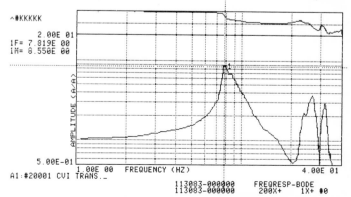

Fig. 28.7 Transfer characterisitics of an electrical cabinet. Courtesy Farwell & Hendricks.

effect analysis. Consider a particular SRS, the Required Response Spectrum (RRS) of Fig. 28.6. If our f_ns (ranging say 1 to 35 Hz) mathematically model a power plant or an equipment, the RRS tells us (in terms of response) what test mechanical input is needed. Perhaps a year before the test, the designer knew how much strength he should provide in elements having various f_ns.

To verify test input from the RRS: vibrating platform accelerometers must develop (in electronic models of the SDoF units) the required motion. Transfer characteristics per Fig. 28.7 aid in improving design.

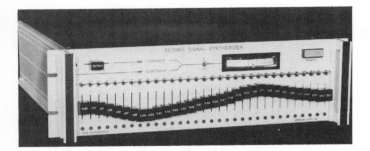

Fig. 28.8 Analog seismic signal synthesizer. Courtesy Bird.

Fig. 28.10 Installation of Earthquake Engineering Reserch Center, University of California. Courtesy MTS.

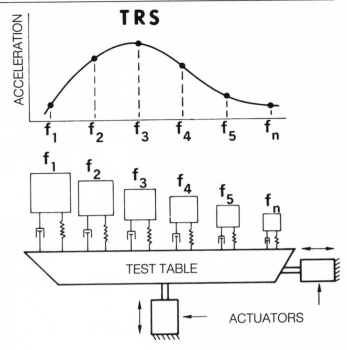

Fig. 28.9 Test Response Spectrum (TRS). After Fischer.

28.4.2 Test techniques of the early 'seventies were directed toward bursts of single-f_f discrete sinusoids and simultaneous f_f combinations. Some laboratories now duplicate magnetic tape recorded earthquakes, or synthesized earthquake tapes, per Sec. 33.4.3. Others manually (Fig. 28.8) or automatically synthesize relatively broad (typically 1 to 35 Hz) spectrum random vibration, comparable to Sec. 25 and 26, appropriately shaped; some add relatively narrow (typically 5 to 7 Hz) random (RoR - Sec. 26.5.2).

To meet an RRS, attach your equipment to a vibrating platform. See Fig. 28.9 and .10. Drive the platform in three axes by three independent random signals into three sets of shakers. Cheaper but less satisfactory to "tilt" a single EH shaker so loads simultaneously receive phase-locked vertical and horizontal motion.

What shaker input is needed to properly simulate earth motions? Most tests today use random signals, but EH shakers can also reproduce the older sine, sine burst, and sine mixtures.

Sincere thanks to Guy Marney for helping with Sec. 28.2.

Section 29
Torsional Vibration

Conventional

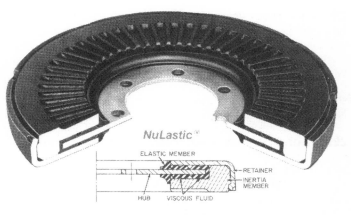

NuLastic

Tee Loc

29.0 Introduction Most of this text deals with vibration assumed (often wrongly) to act in straight lines. (Even EH and EM shakers, for all their motion restraints, develop some lateral and torsional motion.) Many vibration problems originate with rotating machinery (gears, shafts, couplings, impellers, blades, etc.), where forces and responses are often torsional. We here briefly discuss the most common torsional vibration problems and instruments.

Just as with uniaxial vibration, damage rarely occurs without resonance, without the torsional f_f being very close to one of many torsional f_ns. Such machine speeds are often called torsional critical speeds. What little damping exists is mainly in bearings (particularly sleeve bearings) and in shaft materials. Severe angular displacements and stresses can result. As with uniaxial vibration, shaft torsional vibration has maxima and minima (nodes). Don't locate pickups (see Sec. 29.2) at nodes. The shaft end is usually acceptable.

Synchronous electric motors generate surprisingly severe torsional vibrations on startup. By contrast, gas and steam turbines generate little. Small inaccuracies in the machining or installation of gears can produce damaging torsional vibration as well as noise.

Internal combustion engines have crankshaft torsional vibration problems. The input torque is a series of pulses. At certain critical speeds, a shaft torsional f_n is excited; sustained operation hastens shaft fatigue failure. Torsional absorbing or damping devices on the shaft prolong its life.

Fans and other belt-driven accessories on diesel engines experience much torsional vibration, sometimes to failure, usually at torsional resonant peaks. Strobe lights have shown 100-400 Hz peaks from belt "flapping" with nearly one inch D. Tuned absorbers (see Sec. 14.4.3) increase accessory as well as crankshaft life.

29.1 The specialized language of torsional vibration is quite similar to that of earlier sections on uniaxial vibration. See Table 29.1.

Fig. 29.1 Three modern torsional dynamic absorber designs. These employ rubber elements + viscous (silicone oil) damping + a torsional spring and a seismic mass, tuned to absorb vibratory energy at certain f_fs. Courtesy Schwitzer.

Table 29.1

Analogy between uniaxial and torsional vibration

UNIAXIAL SYSTEM		TORSIONAL SYSTEM	
Time	t	Time	t
Mass	M	Polar moment of inertia	$J = Mr^2$
		Radius of gyration	r
Displacement	x	Angular displacement	θ
Velocity	$\dot{x}$	Angular velocity	$\dot{\theta}$
Acceleration	$\ddot{x}$	Angular acceleration	$\ddot{\theta}$
Force	$F = M\ddot{x}$	Torque	$T = J\ddot{\theta}$
Stiffness	K	Torsional spring stiffness	K
Damping constant	C	Damping constant	C
Undamped natural frequency f_n	$\frac{1}{2\pi}\sqrt{\frac{K}{M}}$	Undamped natural frequency f_n	$\frac{1}{2\pi}\sqrt{\frac{K}{J}}$

29.2 Sensors and systems for measuring torsional vibration resemble those for uniaxial vibration. Torsional vibration on rotating parts cannot be readily sensed. Specialized pickups are required for converting it to electrical signals.

29.2.1 Torsional displacement sensors, like the uniaxial relative x sensors of Sec. 3, emphasize low frequency motions. As in Sec. 4.4, most θ measurements commence with angular velocity $\dot{\theta}$ or angular acceleration $\ddot{\theta}$ sensors, followed by single or double integration. Sensors are placed at the ends of shafts or at other locations (via say a belt drive).

29.2.2 Torsional velocity sensors are exemplified by Fig. 29.2. Pulses can be generated by gear teeth passing an inductive sensor, by light bouncing off reflective stripes, etc. Constant $\dot{\theta}$ gives fixed frequency pulses but varying $\dot{\theta}$ gives varying frequency pulses. Frequency to voltage demodulation then gives two measures: the dc component measures the shaft's average speed; the ac component measures variations (if any) in $\dot{\theta}$ and can be integrated to represent θ.

29.2.3 Torsional acceleration sensors originally employed wire strain gages to sense the torsional force accelerating an inertial mass. PR units (Fig. 29.3) today are much smaller. Some contain microcircuitry for low electrical impedance (can output via slip rings) and high sensitivity. They are used on shafts, drivetrains, motors, engines and turbines. Some are used on machine tools to study cutting chatter and tool breakage. Others on crash test anthropormorphic test dummies to measure head and neck rotation. Accelerometers emphasize high frequencies, unless signals are integrated to $\dot{\theta}$ or θ.

29.2.4 Torsional force (torque) sensors generally use wire or semiconductor gages (see also Sec. 3.2.2) for measuring surface strain of, for example, a shaft. The torsional variable differential transformer (TVDT, similar to the LVDT of Sec. 3.2.1) is used on non-magnetic shafts. Strain is proportional to angular force or torque.

For sensing torque on a non-rotating shaft, neither slip rings nor rotary transformer is needed. A "torque table" reaction unit can support and can measure the static and/or dynamic twist of say a motor.

Fig. 29.2 One of several methods for sensing shaft torsional velocity. A magnetic pickup senses passing gear teeth or notches on a disk. Courtesy Econocruise.

Fig. 29.3 PR angular accelerometer. Courtesy Endevco.

29.2.5 Signals off rotating equipment (automobile wheels, propellers, drive shafts, transmissions, gas and steam turbine rotors, etc.) must pass to non-rotating circuitry and thence to measuring and recording apparatus. The traditional approach, usable in clean, mild environments: slip rings (if space is available and accessible at the shaft end) with fixed brushes. Power to excite sensors also crosses slip rings. Main drawback: electrical noise. Better than carbon brushes: silver graphite brushes riding on silver slip rings. Quieter: sealed mercury rotary contacts. Another approach, signals transformer coupled between concentric coils, some rotating and some stationary; avoid misalignment and runout.

PCM telemetry can be used over a short radio frequency (RF) link if temperatures are modest. Advantages include multiplicity of channels onto one RF link.

For harsher environments, consider equally accurate AM/FM telemetry. Sensors include strain gage bridges, PR and PE devices, thermocouples, etc. Several signals can be multiplexed onto each RF channel. A typical 1 inch³ transmitter package is easy to "tuck away" on the rotor. Keep it close to the axis for least centrifugal acceleration.

Fig. 29.5 Torsional EH shaker. Courtesy MTS.

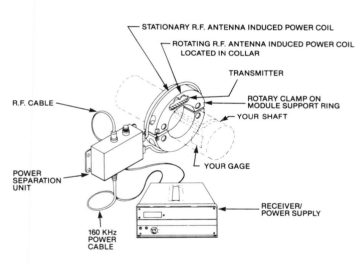

Fig. 29.4 A split collar transmitter clamps around the shaft. The signal is picked up by a stationary antenna, then travels by cable to a nearby radio receiver. Courtesy Acurex.

29.3 Torsional vibration analysis is analogous to uniaxial. See Table 29.1. Signals from the various sensors described in Sec. 29.2 are analyzed as in Sec. 7 and 8.

29.4 Torsional shakers test such units as a diesel engine cooling fan. Fatigue and operational tests demand torsional dynamic forcing over some f_f range, generally using torsional EH shakers (see Fig. 29.5.) They closely resemble the linear EH shakers of Sec. 11. The most useful f_f range is 0-200 Hz, with relatively large θ at low f_fs. They apply dynamic torques to such rotating components as pumps, motors, transmissions, drive trains, vibration isolators, torsional dampers and couplings. θ, $\dot{\theta}$, $\ddot{\theta}$ or strain can be controlled.

Some actuators are rotated by and react against a flywheel, driven by a variable speed electric motor. Differential θ to ±50° (between output head and flywheel) is sensed by a differential θ sensor. Slip rings carry all electrical connections. A disc brake can prevent rotation. Small torsional EM shakers have been used for calibrating rate gyros.

29.5 Torsional slip tables were used for torsional vibration testing, per Fig. 29.6, before torsional shakers per Sec. 29.4. One or two shakers drove a radius extension arm. The test load cg was on the rotation axis, above a hydrostatic bearing.

The authors wish to thank Dr. Ronald L. Eshleman, The Vibration Institute, 101 West 55th Street, Clarendon Hills, Illinois 60514, for his advice and assistance in writing Sec. 29.

Fig. 29.6 Non-rotating torsional excitation by EM shakers driving an oil film slip plate. A torsional hydrostatic bearing is at center. Courtesy STL and Team.

Section 30
Stress Screening

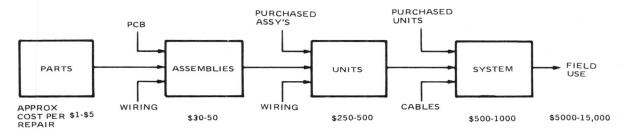

Fig. 30.1 Finding defects at the lowest level of manufacture is least expensive. After Fiorentino and Saari.

30.0 Introduction. Stress screening is a modern electronics production tool. Screening (and subsequent removal from production of flawed units) raises reliability and reduces checkout and field support costs. Initially used on space and military programs, screening now improves such commercial equipment as computers. Screening identifies latent defects (poorly soldered connections, loose hardware, contamination such as bits of solder, wire scraps, etc.) when least costly.

How to best identify flawed assemblies and units? Probing and visual inspection? Microscopic and/or x-ray examination? Temperature shock is faster and more effective than "burn-in." Now add brief intervals of severe broadband (not necessarily random) vibration to simultaneously excite several response modes. Little spectral control is needed.

Flaws may cause difficulty during final checkout, during transportation, during storage or relatively early in service, even in benign environments. The upper curve of Fig. 30.2 applies. Screening causes most failures to occur at the factory. Failed units are removed before further assembly (cuts hidden rework costs); never reach the customer; never require repairs. The center curve applies. Frequent failures demand design and/or production changes to achieve the lower curve.

SCREENING EFFECTIVENESS

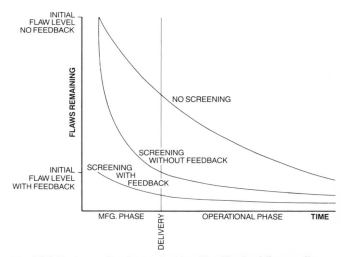

Fig. 30.2 Early portion (upper) of familiar "bathtub" mortality curve. Many "infant" failures are really due to manufacturing defects. Result (center) of screening out defects improves reliability. Even better (lower): improve production processes and/or modify the design. After Baker.

See Table 30.1 and the IES "Proceedings of the Second National Conference and Workshop on Environmental Stress Screening of Electronic Hardware (ESSEH), Sept. 21-25, 1981 at San Jose, California." Also Sept. 11-13, 1984 at Philadelphia. See also the IES "Environmental Stress Screening Guidelines," 1981.

Consider an item intended for a 1g vibration environment. (Multiply g values in this paragraph by your own choice of usage intensity.) Qualification test severity (on a preproduction sample unit) might be 1.5g. If the designer uses *any* severity factor, it will probably be 1.5. Acceptance test severity (on sample production units) might be 1.2g. The long-time reliability demonstration test severity (on sample units) might be 1g. But the screening or *stimulation* intensity on 100% of production might be 6g RMS. That number upsets many designers; many hopelessly try to equate 6g RMS random with 6g RMS sine testing. Show them (Sec. 22.12) that random vibration is comparatively gentle.

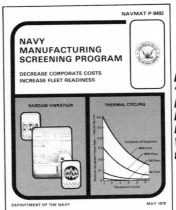

Fig. 30.3 NavMat P-9492, "Navy Manufacturing Screening Program," available from Naval Material Command, Code 06, Room 348, Crystal Plaza #5, Washington, D.C. 20360, telephone 202/692-1106.

Several 1981 San Jose papers, also P-9492 (Fig. 30.3) recommend particular temperature limits and 6g RMS random vibration, per Fig. 30.4. Better: experimentally develop your own screen, one that rapidly precipitates latent defects into hard defects, but doesn't create new defects. Do not try to simulate field use conditions; be severe. Remove any

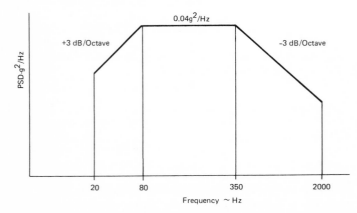

Fig. 30.4 Grumman spectrum from P-9492 is a good "starting point" for tailoring your own screen for your particular product. $\sqrt{\text{area}} = 6g$ RMS.

shock or vibration isolators before screening. "Step stress" one variable at a time on samples having known defects. Yes, you will destroy some samples. Lessen stresses for production screening. Record every failure. Be sure that specific corrective action is taken. Do not worry that many screening failures predict high service failure rates; they may only indicate effective screening. Observe failure modes and mechanisms now and after design improvements.

Most emphasis in Sec. 30 (as in this entire text) is on random vibration, though temperature shock alone is more effective than random alone; see Table 30.1. Start with several rapid temperature cycles per Table 30.2; observe fewer defects with each cycle. (Remove covers to speed internal temperature change.) When the flaw detection rate is sufficiently low, apply random vibration. You might not detect any flaws during vibration. But more defects will appear on the next temperature cycle. Observe additional defects with each cycle, though fewer than during the first few cycles. Combining shaker and chamber saves time over transferring units back and forth.

Table 30.1

Assembly Level Defect Types precipitated by Thermal and Vibration Screens

Defect Type	Thermal Screens	Vibration Screens
Defective Part	X	X
Broken Part	X	X
Improper Installation	X	
Solder Connection	X	X
Faulty Etching	X	X
Loose Contact		X
Wire Insulation	X	
Loose Wire Termination	X	X
Improper Crimp	X	
Contamination	X	
Debris		X
Loose Hardware		X
Mechanical Flaw		X

30.1 Apply power during screening. Heat generated inside components speeds internal temperature rise and stresses. (Possibly disconnect cooling.) Remove power while temperature falls.

Monitor item functioning; watch for any brief "glitches" or intermittent conditions. Otherwise you will miss perhaps 50% of defects.

30.2 Screen at what assembly level? Most screening presently is performed on field replaceable units; see Sec. 30.2.5. Greater benefits can be achieved at lower assembly levels, per Fig. 30.1. Classify flaws at each level which you hope to precipitate, also make a list of failures you might inadvertently cause.

Table 30.2

Possible Temperature/Random Vibration Screening Program and Results

Step 1. Several temperature cycles, during which 1.5, then 0.8, then 0.3 average flaws per subassembly are found.

Step 2. Ten minutes of random vibration, during which 0.02 average flaws per subassembly are found.

Step 3. Several more temperature cycles, during which 1.1, then 0.6, then 0.1 average flaws per sub-assembly are found.

Step 4. Another ten minutes of random vibration, during which 0.007 average flaws per subassembly are found.

Step 5. Several more temperature cycles, during which 0.2, then 0.04, then 0.03 average flaws per sub-assembly are found.

Step 6. Continue alternating random vibration with temperature sweeps until the "fallout rate" approaches zero.

30.2.1 Stress screening of components applies stresses at the part level, where failures are least costly. Two vibration difficulties:

1. exciting the precise flaw location inside the part without creating new flaws, and

2. attaching parts to the shaker (fixturing) at high production rates without damage from handling, especially to pins. See also Sec. 19.

30.2.2 PIND Testing of Components, an alternative to stress screening, refers to Particle Impact Noise Detection tests, usually on sealed microelectronics devices, usually per MIL-STD-833B. Get controversial data from B & W Engineering, Costa Mesa, CA and from Endevco.

30.2.3 Acoustic screening of components. Severe acoustic noise (see Sec. 31) + temperature stressing appears useful for screening ICs. Shakers can seldom excite resonances within small devices. The intervening structures (boxes and PCBs) are too soft to mechanically couple say 5-10 kHz into components. Intense noise bypasses intervening structures to excite resonances.

Suspend devices in a small reverberant temperature chamber; bathe all sides with intense sound. Bouclin (Naval Weapons Center, China Lake) is thus experimentally screening components with temperature shock + 4,000 acoustic watts input.

30.2.4 Boards can carry poorly-soldered connections, wire scraps and bits of solder (which may need to be jarred loose.) Greatest concern: interconnections between and attachment of components. The vibration axis should be perpendicular to the board. Find existing flaws but do not crack the board nor damage good printed circuits nor good board-to-component electrical connections. Apply sufficient stress to find flaws but not fatigue good boards. P-9492's .04 g^2/Hz is rather severe here. Possibly limit first-mode board D to 0.003 $\times$ shortest board dimension. You may need to "notch" such a spectrum as Fig. 30.4, to avoid overdriving one or more modes.

Fig. 30.5 Dual shakers drive fixture which accomodates 7.5 to 18 inch PCBs. Courtesy VTS.

Try to use the same fixture for vibration as with automatic test equipment. Small shaker(s) are generally safe as force limited. Check board responses with a hand-held probe per Fig. 5.33; read D on a meter. Tight spectral control is not required. Be sure that at least the first six bending modes are excited.

30.2.5 Drawers and boxes are now most popular for screening. Assuming per Fig. 30.1 we have already screened parts and boards, here (to say 200 Hz) we mainly check interconnections between subassemblies. This assembly level is convenient for shaker mounting; military drawers and boxes are usually sufficiently stiff. Consider attaching small EM or single or multi-axis pneumatic vibrators per Fig. 30.6 diagonal to hard corners via a relatively soft

Fig. 30.6 Westinghouse "thumpers" diagonally excite boards, whose responses are being monitored. Boards largely ignore spectrum changes.

mounting. Good design (of the unit) insures that its major f_ns (and thus the screening f_fs) are under half the first board f_n, per Steinberg's octave rule; resonance "stacking" could otherwise overdrive boards. Perform a resonance search (Sec. 17.2.1) on a new unit; determine an input that gives modest board responses. (Boards have already been screened.) When Silver et al (Westinghouse) replaced the pneumatic vibrators of Fig. 30.6 with EM shakers, and moderately shaped the spectrum, PCB responses were little affected. So why not use the cheaper pneumatic shakers? Note the overall RMS vibration level; maintain this ±2dB on future screens.

30.2.6 Equipment racks and large boxes should be similarly low frequency screened for faulty connections between drawers or subassemblies. Major rack f_ns should be under half the first drawer f_n. Internal responses will resemble those found in common carrier truck transport; consider screening with a transportation simulator per Sec. 11.5.

30.3 Types of shakers. The mechanical shakers of Sec. 10 are useless here. Consider EM (Sec. 12) and EH (Sec. 11) shakers. Mechanical and pneumatic devices avoid PA (Sec. 13) and controller (Sec. 24-27) costs.

30.3.1 Need for multiple frequencies. We need wide-band excitation (typically to 2,000 Hz for PCBs and small boxes), but it need not be random. No particular spectral shape nor tight tolerances per Sec. 25 and 26 is needed. Shake perpendicular to PCBs to simultaneously excite several response modes. Avoid shaking a PCB via its housing or a handling fixture; neither effectively transmits high f_fs.

30.3.2 Conventional electromagnetic shakers per Sec. 12 are most used for screening to 2,000 Hz, 1979-present. But controls are expensive. P-9492 describes inexpensive audio cassette playback; see Sec. 24.1.4.

30.3.3 Two-axis electromagnetic shakers (see Fig. 30.7) saves time changing specimen axis when boards are not parallel. They also precipitate flaws that remain hidden with single axis shaking. See Sec. 28.4 on seismic simulation, also Sec. 30.3.6.

30.3.4 Conventional electrohydraulic shakers seem effective for screening such large assemblies as terminals and transport/handling racks (at low f_fs and large D). See Sec. 11.

Fig. 30.7 Orthogonal EM shakers, independently (two power amplifiers) driving a slip plate (Section 20.3.2). Courtesy Environmental Screening Technology, Inc.

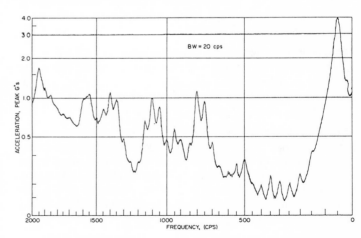

Fig. 30.8 20 Hz spectral analysis of "Rotocon" machine.

30.3.5 Mechanical devices. Various mechanical devices include the General Dynamics, Pomona, CA "Hogan Clanker." Other laboratories instead attach a "bucket of bolts" to greatly distort sine vibration.

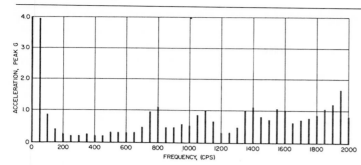

Fig. 30.9 Theoretical line spectral analaysis from Fig. 30.8.

The $10,000 1976 LAB (now Ling/LAB/MTI) "grinder" was described by Noland and Cox, "Low Cost Mechanical Random Shakers," in the 1977 IES PROCEEDINGS, for tests per MIL-STD-781C. Rubbing action yields a broad spectrum with about 3g RMS below 2,000 Hz.

Finally, the Convair "Rotocon" machine was described in "The Development of a Low-Cost Complex-Wave Machine for Quality Control Vibration Tests," by Peterson, SHOCK & VIBRATION BULLETIN 27, part 3. Shaft unbalance at 3,000

RPM gave a spectral line at 50 Hz. "Clanker" assemblies added other lines to at least 10 kHz, resulting in the spectra of Fig. 30.8 and 30.9. Line spectra are effective if they simultaneously excite major resonances. See also "Vibration Testing as a QA/QC Tool," by Lieberman in June 1968, TEST ENGINEERING AND MANAGEMENT. His firm sold many "Rotocons" (under $5,000) during the 1960s under Convair license.

30.3.6 Pneumatic vibrators, inexpensive and reliable pneumatic bin vibrators or "thumpers," generate spectra for stress screening per Fig. 30.6. See "A Low Cost Pneumatic Vibration System," by Silver, Szymkowiak and Caruso in the 1980 IES PROCEEDINGS. Thumper stimulation above 10 kHz can be truncated to 2,000 Hz by elastomers. Further truncation, to say 200 Hz, screens drawers or racks. Avionics board responses resemble their responses to aircraft gunfire, suggesting better environmental tests than Sec. 26.5.3.

We credit Branford Vibrator Co., New Britain, CT, who pioneered pneumatic vibration (Fig. 30.10) to electronics manufacturers. Users find it much cheaper to find and repair defects before shipment.

Another pioneer was General Dynamics (see Van de Griff, Ayers and Maloney, "Simulating Tactical Missile Flight Vibration with Pneumatic Vibrators," SHOCK & VIBRATION BULLETIN 46, Part 3). They modulated air flow and thus "smeared" line spectra into continuous spectra.

The pneumatic vibrator system in Fig. 30.11 digitally modulates air flow for smearing. The resulting spectrum is quite ragged, resembling in-flight vibration spectra. (The spectra of Sec. 23.7 and Fig. 30.4 are overly smooth.) Multiple-axis shaking (see Sec. 20.6) here saves much screening time *and* precipitates flaws that single-axis shaking misses.

Fig. 30.10 An early preumatic vibrator checks a card reader, seeking any defects. Courtesy IBM and Branford Vibrator.

Fig. 30.11 Coauthor Mercado attaching fixture, pre-loaded with 12 subassemblies, to multi-axis pneumatic vibration table inside temperature chamber. Left rack functionally checks 12 units being screened. Right rack controls temperature and vibration. Courtesy Hughes/Santa Barbara Research Center and Screening Systems.

30.4 Spares and repairs are to be screened, per the US Navy Ships Parts Control Center at Mechanicsburg, PA. Replacement and repaired boards will be screened per Sec. 30.2.4. Replacement and repaired boxes will be screened per Sec. 30.2.5. See R. A. Steiner's 1983 Reliability & Maintainability Symposium paper "A Simple Method to Reduce Avionics Repair Costs"; he reports increased reliability from screening elderly electronics boxes.

Electronics maintenance personnel are painfully aware that equipment difficulties commonly "hide" during troubleshooting; service reports commonly report "nothing found." Silver, et al urge combining pneumatic vibrators + inexpensive temperature shock equipment (as in production screening) on electronics troubleshooting benches. "Glitches" that then appear (monitor functions per Sec. 30.1) will reveal that problems exist, that the unit must not be returned to service.

Section 31
Acoustic (Intense Noise) Environmental Testing

Fig. 31.1 Intense acoustic pressures from the engines during lift-off (and from turbulent flow over the skin during transonic flight) stimulate mechanical responses.

31.0 Introduction. Intense noise is an engineering problem aboard high performance jet aircraft, rockets and missiles. When sound pressures exceed 130dB (see Table 31.1), they mechanically vibrate thin parts such as PCBs. Pressures over 170dB have been observed, but 150dB is more common. Spectra are broad, ranging from a few hertz to 10 or 20kHz or higher, so that many resonant modes are simultaneously excited. A brief discussion of intense noise tests (appropriate in situations where intense environmental noise creates problems) follows. This discussion supplements earlier discussions of random vibration testing (appropriate where intense environmental mechanically-transmitted vibration creates problems).

Table 31.1

$$\text{number of dB} = 20 \log \frac{\text{no. of Pa}}{0.00002\text{Pa}}$$

0.00002Pa =	0 dB	Threshold of hearing
.0002Pa =	20 dB	Recording studio
.002Pa =	40 dB	Quiet library
.02Pa =	60 dB	Conversational speech
.2Pa =	80 dB	Typical factory
2Pa =	100 dB	Symphony orchestra
20Pa =	120 dB	Aircraft takeoff
200Pa =	140 dB	Pain
2,000Pa =	160 dB	Rocketry k 0.3 lb/in²
20,000Pa =	180 dB	Valve throat k 2 lb/in²

Acoustic environmental laboratory tests expose components, assemblies, systems or entire vehicles to intense noise to identify possible failures. Specimen responses and any damage should closely resemble those experienced during flight.

31.1 When is acoustic testing appropriate? On aircraft structures; fatigue failures occur close to jet engine exhausts. Also on such critical surfaces as fins, inlet ducts, etc. Acoustic energy excites response modes. Tests lead to improved designs. Both sine and complex excitation (single and multiple sirens) were once used. Other experiments determine noise transmission to predict aircraft internal sound levels. Acoustic cleaning of spacecraft shakes loose any debris. Factory acoustic stimulation may one day stimulate failures in microcircuits, as random vibration is now used on say PCBs (see Sec. 30).

However, Sec. 31 mainly discusses intense noise environmental qualification tests of equipments, similar in purpose to the vibration environmental tests of Sec. 16 and 23. Note that the transmission medium is air, rather than structure.

MIL-STD-810D, Method 515.3 calls for acoustic noise tests to "measure how well a piece of equipment will withstand or operate in intense acoustic noise fields. The acoustic noise test complements tests for structure-borne vibrations." The Standard points out that "In an acoustic noise field, pressure fluctuations impinge directly on the equipment. The attenuation effects of mechanical transmission are missing and the response of the equipment can be significantly greater. Further, components which are effectively isolated from mechanical transmission will be excited directly." Examples of acoustically induced problems are listed: (a) failure of microelectronic component lead wires, (b) chafing of wires, (c) cracking of PCBs, (d) malfunction/failure of waveguides, Klystron tubes, (e) vibration of optical elements. This Method particularly applies to internally-carried airborne equipment, to assembled externally-carried stores, and to ground support equipment on the flightline if levels exceed 135dB or if vibration testing is impractical.

It does not apply to high density (particularly encapsulated) modules, transformers, etc. or equipment surrounded by heavy metal cases, (particularly if potted). The only way to *know* if an item is sensitive to its service noise environment is to conduct a noise test. Many tests are quite short, on the order of one minute. Such difficulties as loose plugs or connections, faulty diodes, etc. show up only briefly, as "glitches."

See also Test Procedure Fg on Acoustic Vibration, part of IEC Publication 68-1 "Basic Environmental Testing Procedures."

Fig. 31.2 Mechanical "point" excitation with an EM shaker aimed at the cg of an aircraft external pod. Internal responses under test will not resemble those of flight, where most responses are caused by turbulent air flow over the entire skin surface. Responses to acoustic tests are much more realistic.

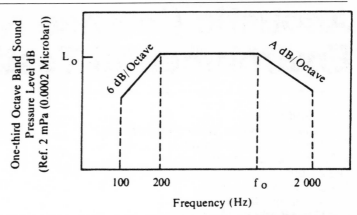

Fig. 31.3 One of several "default" spectra from Method 515.3 of MIL-STD-810D. They are only to be used when environmental data is not available.

31.2 Acoustic environmental test systems have featured jet engine exhausts. However, heat was a problem. As with air jets, spectral content could not readily be adjusted. A few laboratories have used up to 35 inexpensive sirens to combine warbling sounds into a relatively uncontrolled but broad spectrum (FM sirens). Systems amenable to electronic control today achieve desired sound pressures and spectral content. Let us examine valves, controls and chambers. See Fig. 31.4.

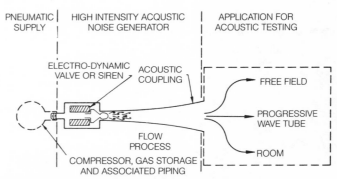

Fig. 31.4 Intense noise originates in flow from a compressed gas tank or more commonly from an air compressor. A valve (EM or EH) modulates gas flow, feeding the test site through a horn. Courtesy Ling Electronics.

31.2.1 Valves generate noise. The acoustic pressure in the horn where it joins the chamber is typically 170 dB. However, at the valve the pressure might exceed 184 dB. At such high acoustic pressures there is much harmonic distortion. Even when the valve produces sinusoidal pressures, in the horn they resemble a sawtooth due to nonlinear behavior of sound waves at high levels. This develops high frequency acoustic content above the maximum valve control frequency. With some valves the amplitude probability density function is quite Gaussian, similar to Fig. 22.6.

Moderate pressure nitrogen gas or air flow is modulated or valved to generate the test fields. Fig. 31.5 and 31.6 show an electromagnetic pneumatic valve (EMPV) or electropneumatic transducer (EPT); coil current (from an amplifier) in a magnetic field activates the valve. Flow is modulated up to about 2,000 Hz; gas flow produces higher f_fs, with some degree of spectral control via gas pressure. EMPVs are offered by Ling of Anaheim, CA and Wyle of Huntsville, AL. Section 31.2.2 discusses controls.

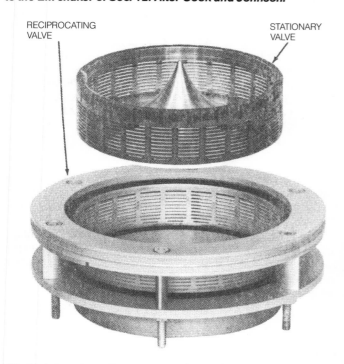

Fig. 31.5 An electromagnetic (EM) valve, similar in principle to the EM shaker of Sec. 12. After Cook and Johnson.

Fig. 31.6 Reciprocating portion (below) of EM valve is axially driven relative to stationary portion (above). Courtesy Ling Electronics.

Alternately, consider electrohydraulic pneumatic (EHPV) or "poppet" valves, said to offer higher efficiency. A reciprocating poppet valve is driven by an EH actuator controlled by a servo valve. Valve peak pressures nearly reach +4 and -1 atmosphere (zero pressure). The shapeable acoustic output is mainly 75-800 Hz + uncontrollable harmonic energy above 800 Hz. Team of South El Monte, CA offers the EHPVs formerly supplied by Northrop. See "A Comparative Evaluation of Three Types of Acoustic Noise Generators," by L. L.

Cook, Jr. and H. B. Johnson, Jr. in the PROCEEDINGS, 1974 IES ANNUAL MEETING.

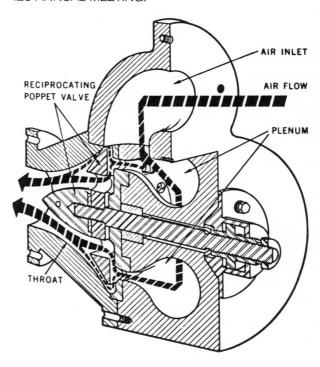

Fig. 31.7 An electrohydraulic (EH) valve, similar in principle to the EH shaker of Sec. 11. After Cook and Johnson.

Fig. 31.8 Noraircoustic EHPV. Courtesy Northrop Aircraft.

31.2.2 Controls for noise testing resemble those for shaker random vibration. To reproduce a recorded sound, use tape playback. To synthesize random sound, take signals from an analog random noise generator or utilize digital control. B is typically 1/3 octave, with tolerances ±3 dB (±2 dB overall). Most tests are run "open loop," with an operator manually shaping the spectrum, much as in Sec. 25.2.2. Digital shaping, comparable to Sec. 26, is being developed. However, it is difficult to control spatial and time correlation, as in simulating a complete flight history.

A number of microphones (at least 3) are placed at different locations around the test unit so that sound pressure levels as well as spectra can be averaged, measured and controlled with limits typically ±3 dB.

31.2.3 Chambers. In a progressive wave chamber, acoustic energy moves from a source to an absorptive termination *past* an item that in service will be exposed to localized noise sources such as exhausts. By contrast, in a reverberation chamber the walls are highly reflective; little sound is absorbed. Acoustic energy impinges all sides of such test items as missiles.

90 feet^3 is a very small chamber. The largest known to us is 44 × 50 × 86 feet high (13.4 × 15.2 × 26.2 m high), at Lockheed Missile and Space Co., Sunnyvale, CA. Room geometry is chosen for best distribution of standing waves down to the lowest frequency of interest. You need a large chamber if testing to low f_fs, as seen for example with large exhaust nozzles on big space booster rockets. A large chamber needs more acoustic energy input to get desired pressures. A large specimen affects the field. Room volume should be at least 10 × specimen volume, with specimen/wall clearance at least $\lambda/4 = 25$ feet (7.6 m) if attempting $f_f = 10$ Hz. In the example cited above, the lowest test frequency is the 1/3-octave band centered at 25Hz.

The specimen is sometimes resiliently suspended, with $f_n < 25$ Hz or 1/4 of lowest specimen f_n, whichever is lower.

Sound energy from the valve must be impedance matched by a horn (proper flare rate and length) to effectively radiate low frequencies. Odd chamber shapes like pentagons with peaked ceilings are sometimes used. The goal: uniform modal density. To minimize wall absorption, walls should be hard and rigid. The chamber should be massive, with double walls and sound leaks sealed, so energy is not lost and personnel danger is minimized. Gas (N_2 presents safety problems) exhausting from the chamber should be muffled.

31.2.4 Instrumentation. Instrumentation includes external microphones to measure the applied environment, internal microphones to measure the internal environment and skin transmission characteristics, also accelerometers and strain gages to measure the resultant vibratory responses.

31.3 Combined environments. There is much evidence that the only way to reproduce the service failure distribution is to reproduce the service stress distribution (the range of stresses, in proper proportion, of levels and durations measured during typical flights).

Fig. 31.9 shows a test chamber at the U.S. Navy's Pacific Missile Test Center, Point Mugu, California. Missile guidance

Fig. 31.9 Six Sidewinder missile guidance packages are simultaneously exposed to three environments. Courtesy Pacific Missile Test Center.

and control sections (power on and functioning) are tested under combined noise, vibration and thermal shock (-40 to +60°C). Sections under test are mounted on inert rocket motors. Noise enters through the 100 Hz horn opening on facing wall. Heated or cooled air passes through fixed and flexible ducts to a silicone rubber temperature shroud (acoustically transparent). In addition, each pair of weapons (mounted as in service) receives low f_f random vibration from an EH shaker. Input adjustments bring internal temperatures and vibration responses close to those measured during captive carry flight on aircraft. Test failures closely resemble those observed in service; these failures were not achieved until all three environments were applied concurrently. PMTC experience is reflected in Method 523.0 of MIL-STD-810D.

31.4 Acoustic inputs for stress screening seems somewhat better than random mechanical vibration per Sec. 30, particularly for such small items as ICs. Shakers and fixtures are not very practical for stimulating internal resonances at very high f_ns. Some preliminary work is being conducted at Naval Weapons Center, China Lake, CA; we look forward to hearing of experimental results. Lower-cost alternatives include multiple pneumatic vibrators per Sec. 30.1.6.

31.5 Facilities. Interested readers may wish to visit an acoustic test facility. See the 1975 ACOUSTICS TESTING FACILITY SURVEY produced by the IES Acoustics Committee. In 1975 some 33 USA facilities could do high-intensity noise exposure tests. By now there must be 100. Hiring an existing laboratory to perform tests is, as with vibration and shock testing, a good way to commence.

We greatly appreciate the help of G. H. Bosco, H. McGregor and D. C. Skilling in preparing Sec. 31.

Section 32
Shock Measurement, Calibration, Analysis and Design

32.0 Introduction — what is shock? Dictionaries variously define "shock," but this seems useful: a relatively brief (compared to the lowest natural period of involved structures) impact or collision. Decaying oscillations (ringing) of the structures follow.

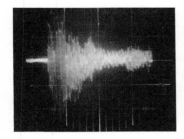

Fig. 32.1 A time history ('scope picture) of a typical structural response to a typical shock. It might result from railway "humping." or, on a vastly shorter time scale, from an explosion.

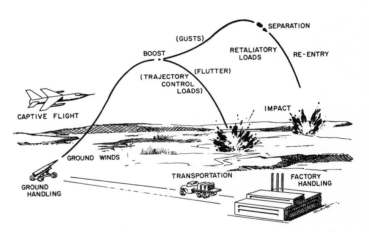

Fig. 32.2 Various dynamic events in a weapon's life, with no clear division between vibration (long term) and shock (short term).

Vast sums are lost each year during goods transport, especially during loading and unloading. Shippers and carriers should be concerned. More engineering attention has been paid to military and aerospace (particularly pyrotechnic — explosive) shocks than to cargo handling.

Engineers often wish to quantize (measure and express in numbers) peak shock amplitudes A as m/s² or g units. Having done so, they feel ready for engineering decisions. But A alone seldom measures damage potential. Frequency content and duration T of a shock pulse are more useful.

Sec. 32.1 considers mechanical shock measurement devices. Sec. 32.2 shows that accelerometer systems (Sec. 5) can measure shock *if* the resulting information is properly processed and stored (Sec. 32.3). Systems must be calibrated, per Sec. 32.4. But shocks are complex; a single number won't express damage potential; Sec. 32.5 and .6 discuss spectral content. Sec. 32.7 discusses shock for modal tests.

32.1 Mechanical devices not satisfactory. In the authors' opinion, purely mechanical shock measurement devices are inaccurate. These cheap devices are alluring: no electrical power, for example. Most are purchased for installation in cargo. Accelerometer systems per Sec. 5 better measure $\ddot{x}$ over a wide T and frequency range.

32.1.1 Simple tripping devices exist in wide variety. Fig. 32.3 suggests steel balls separated by compression springs. They *should* dislodge if $\ddot{x}$ exceeds the rating. Look to see (or shake the cargo) whether the device has tripped. You would need many such devices, variously rated, variously located, to quantize the greatest shock during a journey.

But can a simple spring-mass device accurately quantize shock? Before you buy *any* instrument, inquire into its manufacturer's calibration. Simple static centrifuge calibration is not enough. Shock calibration per Sec. 32.4 is needed. See

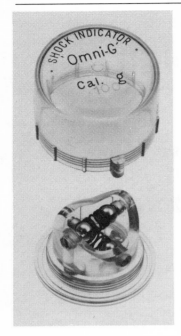

Fig. 32.3 Inexpensive spring-mass device to be attached to cargo. Tripping is supposed to indicate that 100g has been exceeded. Courtesy Impact-o-Graph.

"Analysis of Shock Indicator Devices for Packaging Applications" by R. V. Brown, DSPT Report 73-52, Air Force Packaging Agency Report 71-30-D-54, July 1973. Also his DSPT Report 75-76 "Monitor Fast Pack (Type 1) Shipping Environment," December 1975.

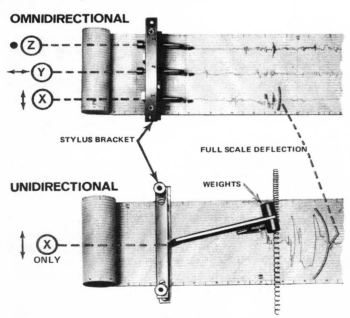

Fig. 32.4 Spring-mass shock recording device attaches to cargo. Deflections are supposed to indicate ẍ. Courtesy Impact-o-Graph.

32.1.2 Mechanical recording devices. No simple device tells when shocks took place (whether in our shipping department, en route or in our customer's receiving department) nor the magnitude nor number of lesser shocks that occurred. Recorders similar to Fig. 32.4 indicate something happened in the x, y and z directions. You can infer times

and dates. Operating without outside power is attractive for attachment (open or secret) to cargo. Shippers seeking damages from carriers often use these "measurements" in their claims, and carriers seldom protest. The authors feel shippers would have great difficulty demonstrating instrument accuracy.

Since mechanical spring-mass systems are frequency sensitive, they respond most strongly to dynamic inputs whose f_rs or T produce pickup resonance. Without calibration against standards (see Sec. 9 and 32.4), you cannot trust the "numbers" generated by those systems.

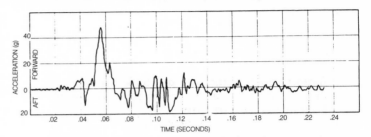

Fig. 32.5 Longitudinal ẍ vs. time during 10 mph rail car "humping." Such data not obtainable with devices of Figs. 32.3 and .4; sophisticated instrumentation needed.

32.2 Accelerometer systems preferable. Accelerometers per Sec. 5 measure shocks and record them for later readout, perhaps as in Fig. 32.5. With less time involved, shock is more difficult to measure than is vibration.

Visualize, lacking a device per Fig. 32.6, trying to use accelerometer systems in cargo monitoring. At this moment cargo is moving on a fork-lift truck. A cable runs to an instrumentation van moving alongside. Will the driver behave normally? No. Better to telemeter signals to a receiver and recorder. Or place batteries and recorder inside the cargo. For realistic data, handlers must not know of investigation.

32.2.1 High-frequency response of accelerometers much concerns us. Consider Fig. 32.7. The sensors were (a) 35,000 Hz f_s lightly damped PE, (b) 2,000 and (c) 850 Hz f_s damped strain gage. We recommend (a) because the low (b) and (c) f_s might obscure significant structural motion. But how much of (a) is structural ringing, and how much is accelerometer ringing?

32.2.2 Low-pass filtering can compensate for "ringing" and simplify explaining records. Deciding upon filter f_u is crucial. Set it low enough to block ringing, say at half f_s. But not so low as to lose useful structural data. Your decision is eased with high f_s accelerometers. Rarely is information needed above 10 kHz, a good f_u when f_s is 50 kHz or higher.

32.2.3 Mechanical filtering. Consider mounting your accelerometer on a pad that passes structural frequencies but blocks "ringing" frequencies. Pad selection is eased by a high accelerometer f_s. Insist upon a proper shock calibration (Sec. 32.4) *with* that pad.

32.2.4 Measuring pyrotechnic shock. These result from explosive discharges, as from explosive bolts. Structural response is very oscillatory, with about equal + and − peaks.

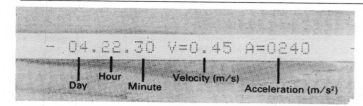

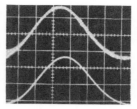 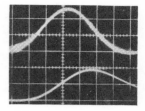

Fig. 32.8 Experiments with setting filter f_u. Upper traces show unfiltered records of empty shock test machine impacting rubber cushion, duration 0.011 sec. Lower trace (left) with $f_u = 600$ Hz (15× lowest f_t); (right) $f_u = 60$ Hz (1.5× lowest f_t) per a common "rule of thumb". Note pulse distortion and delay. Courtesy L.A.B.

strain rates. Pyro shocks cause failures in fragile or brittle components or those having high f_ns, above say 8,000 Hz; vibration tests do not prove shock survivability. See Sec. 33.3 and .4 on pyro shock testing.

All shocks are difficult to measure accurately; pyro events are most difficult. Shocks are very short (event over in 0.02 sec. or less), while peaks are very high (over 1,000g), with most energy above 1,000 Hz. Special "shock" accelerometers are needed: very high limits and f_ss. The PR unit of Fig. 32.9 seems a significant breakthrough. Most shock accelerometers have been PE.

Avoid amplifier overloads. When an amplifier saturates, it temporarily ceases to operate linearly and may lose important parts of the signal. The resulting time history and spectra are incorrect. A low pass filter (Fig. 32.10) before any amplifier helps protect against overload. Amplifiers must be able to deliver adequate voltage and current to their loads, including the surprisingly large capacitance of long output signal lines and large current needed for adequate "slew rate" in readout instruments.

Check your entire shock instrumentation system (including your magnetic tape recorders) to be sure all units faithfully amplify and store "pyro" signals. Some recorders, amplifiers, detectors and readout devices are acceptable on sine signals but fail miserably with short, severe pulses. Shock calibrate the system per Sec. 32.4.

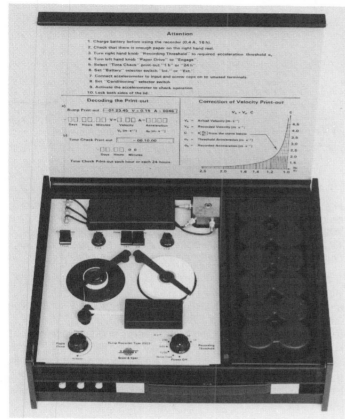

Fig. 32.6 "Bump Recorder" combines three accelerometer signals into numbers representing $\dot{x}$ and $\ddot{x}$, printing onto paper tape. Courtesy B & K.

Cracks don't have time to grow. Often only one part of a specimen has large stresses at a time. Relays may chatter or transfer. Brittle materials may shatter. Structural failures are rare; yield and ultimate strengths are often high at these high

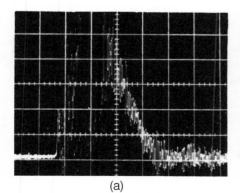 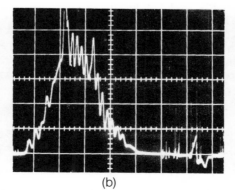 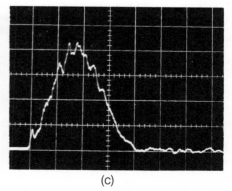

(a)   (b)   (c)

Fig. 32.7 Three oscilloscope records ($\ddot{x}$ vs. time) of the same shock event, sensed by three different accelerometers. Which record is correct?

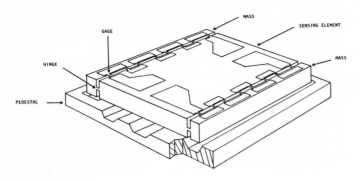

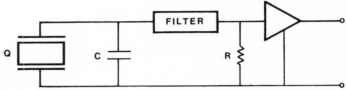

Fig. 32.9 Recent 7270 PR shock accelerometer, range exceeding 100,000g and mounted f_s around 1 mHz. 1 mm square silicon chip is spring + mass + PR gages. Courtesy Endevco.

Fig. 32.10 305MXX quartz pyrotechnic or "pyro" PE accelerometer. Internal amplifier ranged for 20,000g, available to 100,000g. Active two-pole Butterworth low-pass filter flat to 10,000 Hz. Note soldered connections (more rugged than disconnect) and integral mounting stud. Courtesy PCB.

should extend to zero Hz or dc. Strain gage or PR accelerometers per Sec. 5.4 and 5.5 are somewhat better here. Only recently have PE accelerometers been used below say 20 Hz, per Sec. 5.10 and 5.16. Traditionally we have blocked signals below (but passed signals above) our lowest frequency of interest, say 1 Hz. Always use the dc (zero frequency) input to your 'scope; the ac input cuts off signals below about 5 Hz.

Ideally, sensitivity (mv/g or pc/g) should be constant under all measurement conditions. Unfortunately, all sensors are somewhat nonlinear. This problem is least with "shock" accelerometers, designed for relatively low dynamic stresses. PR accelerometers shift zero less and are better for single or double integration per Sec. 3.2.3. However, PR units can break if f_s is severely excited.

Nonlinearity may occur in sensors but more frequently in amplifiers. Data is never completely certain before a measurement (else why perform it?) Data analysis efforts are fruitless if any part of the measuring system saturates. New frequencies (that were not present in the physical motion) are created in the analysis, possibly masking the data being sought. Select the proper range so your electronics do not saturate. Place filters early in the electronic "path."

Zero shift is any analog signal which does not average zero after the motion stops. Cause may be in either electronics or pickup, but don't confuse it with undershoot due to poor low frequency response per Sec. 5.10. Carefully attach cable to connector; best, solder cable directly to the sensor.

32.2.5 Low frequency response is also important, especially with long duration shocks. In theory, your frequency response

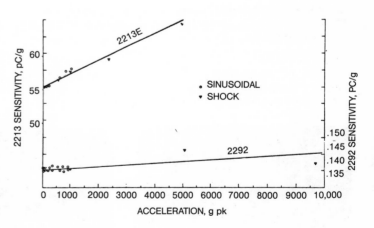

Fig. 32.11 These PE accelerometer sensitivities increase with shock intensity. Effect is least with "shock" unit. Courtesy Endevco.

32.3 Digital transient recorders (DTRs) — time domain. We value sensors (such as accelerometers) over devices

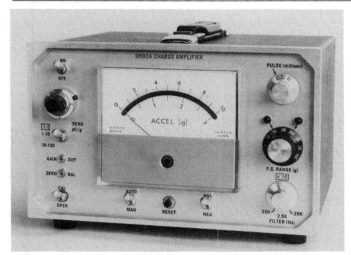

Fig. 32.12 Analog shock-indicating meter. Pointer holds peak indication until operator presses "Reset" button. Courtesy U-D.

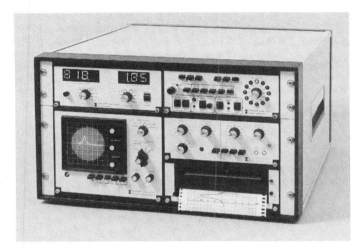

Fig. 32.13 Transient recorder. The digital A and T readouts are only useful on classical pulses. They or "real world" oscillatory pulses can be stored digitally, processed, displayed on CRT and plotted on strip chart. Courtesy Endevco.

per Sec. 32.1 partly because we can store information on magnetic tape (digital or analog). Playback recreates our data. If, say, we've recorded cargo motion during a journey, we can obtain number of peaks, A, dates, etc.

Special analog "shock" meters per Fig. 32.12 hold peak A (really only useful on older, simple "classical" pulses per Sec. 33.2) until manually reset. Digital units per Fig. 32.13 display and record time history.

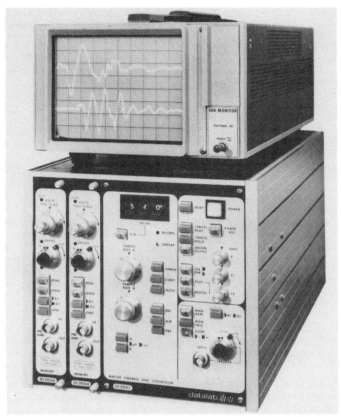

Fig. 32.15 DTR upgrades conventional 'scope. Courtesy Datamation and Transamerica Delaval.

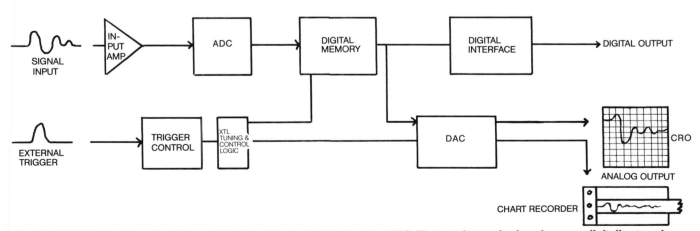

Fig. 32.14 DTR incoming signal is amplified (or attentuated) before the ADC. Time and magnitude values are digitally stored and/or output to a computer. They are also DAC'd to a conventional 'scope (continually refreshed for a flicker-free display), o'graph, chart recorder and magnetic tape or disk recorder.

DTRs extend the shock measurement capabilities of conventional analog time domain instruments: oscilloscopes (Fig. 32.15), oscillographs and magnetic tape recorders. With a DTR, a conventional 'scope can outperform a storage 'scope, permitting positioning and expanding the trace. Long-time events such as earthquakes (Sec. 28.3) can be speeded up. Very short "pyro" shock events (Sec. 32.2.4) can be slowed down. With a DTR, an oscillograph, even a pen recorder, can trace and expand parts of microsecond (and longer) shock events. The DTR's internal digital signal may also feed a computer per Fig. 32.16 for shock recording and analysis; the DTR is often found in digital 'scopes, possibly with digital integration to $\dot{x}$ or to x.

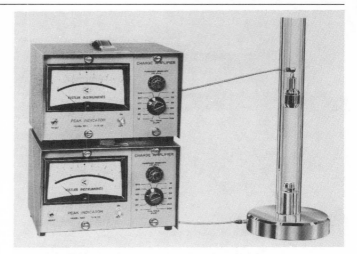

Fig. 32.17 Shock calibrating a falling accelerometer when its carriage impacts a stationary force sensor. Courtesy Kistler.

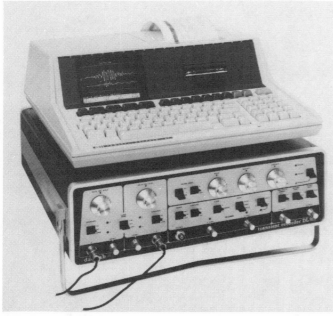

Fig. 32.16 A DTR greatly improves digital recording of shock events, particularly if (like earthquakes) lacking a convenient trigger pulse. Courtesy Transamerica Delaval and H-P.

To record triggerless events such as earthquakes, use self-triggering. Events before and after the main pulse are available for viewing; position the main pulse anywhere on your 'scope screen. A multi-channel DTR permits simultaneous recording of shock inputs and responses at several structural locations; this is particularly valued when a shock cannot be repeated.

32.4 Shock calibration of accelerometer systems (measurement, storage and analysis instruments) at anticipated intensities and pulse durations is vital. Vibration calibration on a shaker seldom exceeds 150g. But what is the sensitivity at 1,000g? At 10,000g? See Fig. 32.11.

At a single f_f, a resonant beam per Sec. 19.14 will increase your shaker's A to perhaps 1,000g. Shock calibration suggested by Fig. 32.17 checks linearity to thousands of g, at various T. Or consider the "Hopkinson bar" technique discussed at the 1983 ISA meeting by Sill of Endevco.

See Corelli, "Ratio Calibration - the Right Choice for Modal Analysis" in SOUND AND VIBRATION, January, 1984. He

describes simple 1g gravimetric step calibrations of accelerometers, force sensors and modal impact hammers. Or write to Dave Corelli at Box 19299, Cincinnnati, OH 45219.

Your oscilloscope + camera records pulse height, shape and duration. A DTR per Sec. 32.3 is helpful. See "The Absolute Calibration of Pickups on a Drop-Ball Shock Machine of the Ballistic Type," by R.R. Bouche, available from Endevco.) Sec. 9.7.3 describes fast digital processing.

Authorities differ regarding attainable accuracies, but agree that shock measurements are less precise than vibration. Accuracy is possibly ±1-2% with vibration vs. possibly ±5% with shock in the standards laboratory. Double those numbers in the field.

32.5 Fourier analysis of shock. We've learned to quantize pulse height, duration and shape in the time domain. We more often need the energy content of shock events at various f_fs; this affects structural responses at various f_ns. See Sec. I-3.3 of MIL-STD-810D re the maximax spectrum.

We saw in Sec. 7.10 how the FFT converts from the time domain (instantaneous force, x, $\dot{x}$ or $\ddot{x}$ vs. time) into the frequency domain (force, x, $\dot{x}$ or $\ddot{x}$ vs. frequency). With a frequency description of a probable service shock, a designer knows what his f_ns should be, for least damage.

Don't use ASD/PSD plots for shock; they can lead to the erroneous conclusion that shock is equivalent to low level random vibration.

32.6 Shock response spectrum (SRS) is another frequency analysis. (See background note in Sec. 28.4.1.) Visualize several SDoF systems per Fig. 32.18, attached to a structure. Each has a different f_n, but the same Q. One of these is shown in Fig. 32.19. A shock pulse occurs; its x (displacement input time history) is pen recorded. Another pen traces y, the response motion. Some masses respond vigorously, due to more input energy at certain f_fs. We observe the maximum response of each reed and plot this vs. SDoF f_n; this graph is called a *shock response spectrum* (SRS). It suggests optimum equipment f_ns: where energy is least. SRS does not

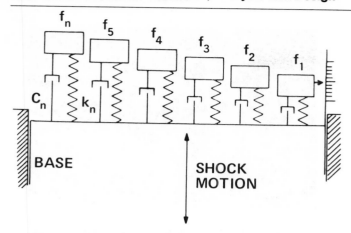

Fig. 32.18 An imaginary series of SDoF systems per SRS concept. Their individual responses to a mechanical shock represent real structure responses to shock.

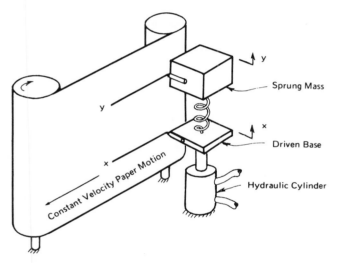

Fig. 32.19 Detail of one SDoF from Fig. 32.18 (damper not shown). x represents displacement of driven base. y represents response of sprung mass.

Fig. 32.20 Navy reed gage (1942), practical implementation of Fig. 32.18, attached to a ship to evaluate shock events. Each reed's x (close to SDoF behavior) is scribed on waxed paper for later measurement.

Fig. 32.21 A modern version of Fig. 32.20. Triaxial units evaluate earthquakes. Courtesy Engdahl.

describe the shock pulse itself; rather the effect of the pulse on a series of SDoFs (representing a real structure). SRS indicates a particular shock's *potential for damage*.

Note that the original pulse cannot be reconstructed from the SRS. Many different time histories can have the same SRS.

SRS are today usually obtained electronically. In analog systems, a pulse from an accelerometer excites an array of contiguous, constant-Q electrical filters (modeled on mechanical SDoF systems) as in Fig. 32.22. The peak voltage from each filter is stored, sequentially sampled and plotted vs. filter f_o, as suggested by Fig. 32.23.

Under digital SRS analysis (sometimes offered as part of shaker control), the operator will have entered accelerometer sensitivity, resolution, B, C/C_c, etc. into his computer. The system is pre-adjusted to trigger when a transient arrives (real time or magnetic tape or disk). Then it is sampled and

displayed in the time domain or in the frequency domain as Fourier spectra, SRS or individual filter responses. To compute the SRS, the pulse is applied to a large number of shock filters, mathematical models of SDoF systems, each having a different f_o. Filter f_o can be changed in 1/3, 1/6, or 1/12

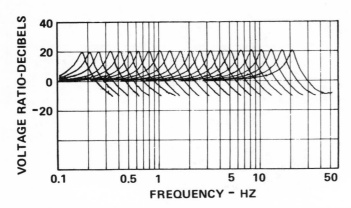

Fig. 32.22 Analog analyzer having 22 filters, here for earthquake SRS.

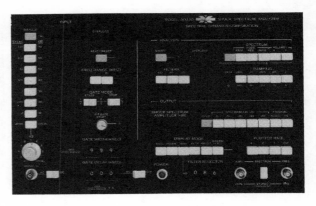

Fig. 32.24 SD320 Shock Spectrum Analyzer stores data in 4,000 word recirculating memory. Analog time or frequency data appears on 'scope or X-Y plotter. Courtesy Scientific-Atlanta.

octave steps. Various spectra such as positive or negative, primary or residual or maximax can be displayed and documented. Spectra can be "tilted" to represent x, ẋ or ẍ. See Fig. 32.24.

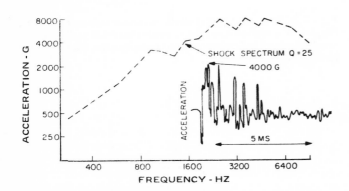

Fig. 32.23 SRS and time history of a pyro event, firing of an explosive bolt. Per Sec. 32.2.4, most energy is at high frequencies.

32.7 Impedance hammers for modal investigations. When the instrumented hammer of Fig. 6.4 strikes a structure, it briefly transmits force. Force is converted to an electrical signal per Sec. 6.1, then recorded digitally, along with response signals from accelerometers and strain gages at various points on the structure.

Data concerning the structure's geometry and dimensions, response sensor locations and a schedule of impact points has already been keyboarded. The computer processes the data from each blow. On demand, it lists structural f_ns and graphically displays modal responses at each f_n.

Unfortunately, ten operators will get ten plots off the same structure; plots may agree at low frequencies, but often disagree at the high end. Operators need considerable training to avoid the many variables. A controlled-force electric hammer, for example, is far more repeatable than any operator.

Section 33
Shock Testing

33.0 Introduction. Sec. 33 considers three types of shock testing specification: 33.1 dictates the machine and how to use it. 33.2 defines the motion desired. 33.3 and 33.4 deal with the "damage potential" of the desired shock (described by the SRS, preferred under MIL-STD-810D). Finally, "miscellaneous" shock tests are discussed in 33.5. Sec. 33 does not deal with air blast nor underwater shock nor nuclear explosions. Earthquakes and explosively-induced earth motions are discussed in Sec. 28.

Equipment must survive (and operate in spite of) service and transport shocks. To prove shock ruggedness, we apply controlled mechanical shocks that (to some degree) simulate field environments. Fortunately, tests need not duplicate field conditions. They need only identify fragile units. Hopefully, our test results will agree with field experience. Generally, we measure field shocks; then we develop a test to simulate the important characteristics of field shocks. Laboratory tests have certain advantages over field experience in determining the shock resistance of hardware: (1) We can control the nature and number of shocks. (2) Our test inputs can represent extreme or average field conditions. A field shock test usually involves only one specific test condition. (3) Easier to instrument and monitor effects. (4) More convenient and less expensive; thus we can develop components and subassemblies, as well as systems, for shock resistance.

Vibration tests generally cannot substitute for shock tests; too often, units that have passed shaker tests fail field shocks. An exception, per MIL-STD-810D, Method 516.3: if 3σ random vibration tests at all f_fs exceed the SRS $Q = 10$ tests specified, the latter can be omitted.

Per Method 516.3 of MIL-STD-810D, "Shock tests are performed to assure that materiel can withstand the relatively infrequent nonrepetitive shocks or transient vibrations encountered in handling, transportation and service environments. Shock tests are also used to measure an item's fragility, so that packaging may be designed to protect it, if necessary, and to test the strength of devices that attach equipment to platforms that can crash."

This text (emphasizing random vibration) deals briefly with some aspects of the nine procedures into which Method 516.3 is divided. (a) Functional Shock (Procedure I); see Sec. 33.3 and 33.4. (b) Equipment to be packaged (Procedure II); see Sec. 33.5. (c) Fragility (Procedure III); see Sec. 33.5. (d) Transit drop (Procedure IV); see Sec. 33.5. (e) Crash hazard (Procedure V); see Sec. 33.3 and 33.4. (f) Bench handling (Procedure VI); see Sec. 33.5. (g) Pyrotechnic shock (Procedure VII); see Sec. 33.3 and 33.4. (h) Rail impact (Procedure VIII); not discussed in 33. (i) Catapult launch/arrested landing (Procedure IX); not discussed in 33.

33.1 Specification describes apparatus. Test engineers sometimes describe a suitable apparatus and how to use it. A wide variety of unduly long-lived machines results. In his 1976 IES annual meeting paper "Equipment Sensitivity to Pyrotechnic Shock," Luhrs compared the effects on a spacecraft structure of a) drop machine half sine pulse, b) shaker oscillatory motion, c) strain energy machine, d) actual pyro on a spacecraft, and e) weight impacting anvil. While their spectra were similar, a) caused more failures. For the same peak response spectra, a) reached much higher peak g, with more low f_f energy. b) and c) had similar time histories and about equal damage potential. Internal response spectra were studied; d) was mildest. Some electronic parts survived all five tests. But relays, for example, were destroyed by some a) and b) tests while not even chattering on e) tests. While not all designers are involved with spacecraft, source mobility per Sec. 6.6 and 6.7 applies to many engineering fields. Designers must understand the shock fragility of their parts and design for the shock environment. If you cannot get sufficiently rugged parts, let a soft structure attenuate the high f_fs; perhaps isolate per Sec. 1.1.5.

We will examine "drop" machines similar to a) in Sec. 33.2.1; oscillatory pulse tests on EM shakers, similar to b) in Sec. 33.2.3, 33.3 and 33.4; strain energy machines similar to c) in Sec. 33.1.3; use of explosives, similar to d) in Sec. 33.1.1; and metal-to-metal impact, similar to e), in Sec. 33.1.2 and 33.1.4.

33.1.1 Explosives — "pyro" shock tests.
Sec. 32.2.4 introduced "pyro" shock and its measurement. Tests determine if equipment will survive explosive service discharges. We should know the field source, the intervening structure (its dynamics at various f_fs and intensities), also the test unit's various failure modes. One way to identify a "good" test: much the same failures in test as in service.

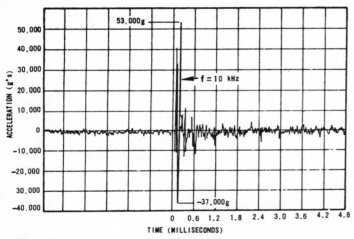

Fig. 33.1 Accelerometer time history close to linear shaped charge which separates missile stages. Note almost instantaneous peak loading. After Powers, 1976.

Generating as in Sec. 33.3 an RRS matching a field SRS (Sec. 32.6) does not guarantee a good test. Explosives fairly well reproduce field pyro shocks; both almost instantly release energy. Nearby sensors (or products) receive a severe short-duration high f_f transient much like Fig. 33.1. However, elsewhere that transient is modified by structural response, similar to Fig. 33.2. Both are pyro events; our test must consider the shock source and nature, also the impedance of the intervening structure (affects source mobility for product *and* sensor.)

Most labs have "classical" drop machines per Sec. 33.2.1. But do not use them for pyro. They greatly overtest at low f_fs; compare their SRS with field events. Pyro features high peaks (commonly 5,000g, as much as 100,000g) and high f_fs. Pyro inputs are oscillatory; classical pulses are not, have dif-

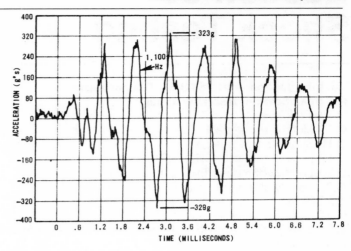

Fig. 33.2 Time history 175 inches from explosion of Fig. 33.1. Note that input is almost completely masked by 1,000 Hz structural resonance. After Powers, 1976.

ferent $\dot{x}$ and $\ddot{x}$. A part's KE absorption is the key to surviving classical pulses (high KE, much low f_f content). Pyro shock (low KE, much high f_f content) generally fails high f_n components.

EM shakers are useful for some low level prescribed spectrum tests, but armature resonances and peak A are limitations. Setup time is minimal. Closely match pulse duration and composition; avoid low f_f overtest. Tests are flawed by unrealistic similarity at all attach points. Real structures (e.g. aircraft) are compliant. Shakers, drop tables and strain energy machines are all rigid; mobility (see Sec. 6.6) is much different. Shakers tend to severely overtest (too-severe internal responses), compared with explosives in service and in testing.

Even if we exactly duplicate an $\ddot{x}$ time history from an actual field event, per Sec. 33.4.3, the loading will differ. Service shocks travel via compliant structure (field and laboratory paths may differ) to and through any attached part.

Moderately tailored explosive "barge" tests are used by various navies; see Sec. 33.1.4. Most spacecraft pyro tests are tailored for a given pyro environment.

Explosives yield high accelerations. But they lack standardization, little control spectral shape (very broad tolerances) and are hazardous.

33.1.2 Metal-metal impact,
as in Fig. 33.3, simulates "pyro" shock. Some laboratories find it easy to control and to repeat. Most test structures (joining anvil and test load) simulate the service structure local stiffness and weight. Varying the hammer blow and the surfaces varies intensity and f_f content somewhat. More energy goes into low f_fs than with explosives (Sec. 32.1.1); also, there is lesser expense and danger. A somewhat different philosophy is found in Navy "hammer" tests per Sec. 33.1.4.

33.1.3 Strain energy release
can generate test "pyro" energy. Strain energy can be cheaply stored in I beams. A coupon ruptures to excite a fixture and test article. Results

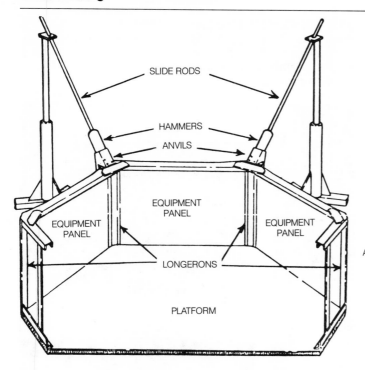

Fig. 33.3 Falling weight impacts anvil, generating shock into sim-ulated spacecraft structure carrying test items. After Luhrs, 1976.

exceeded those obtained on a shaker and synthesizer (discussed later in Sec. 33).

33.1.4 Navy shock tests. Navy shipboard equipments must operate reliably in combat. Certain shocks might not sink a ship, but might disable it. MIL-S-901C (U.S. Navy — 1963) requires specific shock machines and describes their use. See "Procedures for Conducting Shock Tests on Navy Class HI (High Impact) Shock Machines for Lightweight and Mediumweight Equipments," by E. W. Clements, NRL Report 8631, 1982. There is some similarity to guns firing and/or enemy action. Test shocks are considered equivalent if test damage resembles service damage. Tests are verified by tests on entire (surplus) ships.

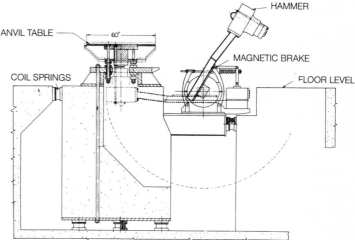

Fig. 33.5 U.S. Navy MWS (medium weight shock) test machine per MIL-S-901C.

Per Fig. 33.4 a 400 lb. hammer drops from specified heights into a vertical 5/8 inch steel plate carrying test items to 400 lb. They experience a multi-f_f decaying transient with peaks to 1,000g. Most energy is 200-2,000 Hz, with peaks near 100 and 350 Hz. Per Fig. 33.5 a 3,000 lb. hammer upwardly strikes an anvil carrying test items to 6,000 lb. Shocks unfortunately change with loading, and peak at anvil f_ns. Two sizes of FSP per Fig. 33.6 carry loads to 60,000 and 400,000

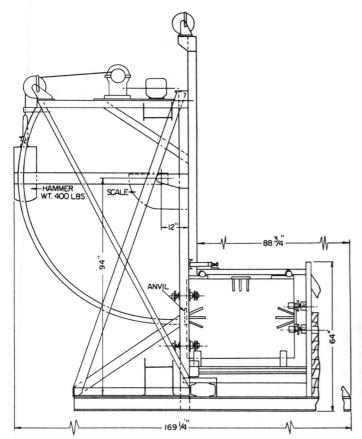

Fig. 33.4 U.S. Navy LWS (light weight shock) test machine per MIL-S-901C.

Fig. 33.6 U.S. Navy Floating Shock Platform.

Ib. Explosives alongside and beneath triaxially excite barge and load. See "The West Coast Shock Facility," by Giannoccolo, MARINE TECHNOLOGY, July 1966. Damage potential equal to actual service — same SRS — is sought.

But how to predict the result on a new test item? Many items fail, undergo redesign, fail again and repeat that cycle until *eventually* they pass. Unfortunately, there is little "learning" to speed future development. Rumors exist that MIL-S-901D will adopt SRS (Sec. 33.3).

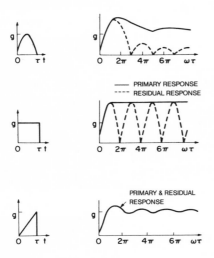

Fig. 33.7 "Classical" pulses (left) that earlier Standards have specified (possibly because easy to describe — certainly never found in the real world). Residual response spectra (right, solid lines) are the loci of maximum responses of systems, attained after the shock ends. Primary response spectra (broken lines) are the loci of maximum responses of systems, attained during the shock. Maximax responses (not shown) follow either the primary or residual, whichever is greater at each fₙ.
Sawtooth pulses (see Fig. 33.8) are valued for non-symmetry and flatter spectrum.

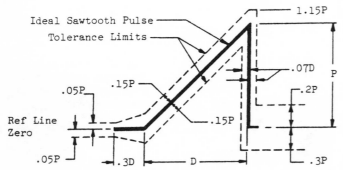

Fig. 33.8 Per method 516.3 of MIL-STD-810D, a sawtooth with peaks to 75g and durations to .011 second. Compared to other classical pulses, its spectrum is relatively flat; see Fig. 33.7. + and - SRS are identical; we can do 6-axis test with 3 impacts. This is a "default" test for labs not capable of SRS. Trapezoidal pulses are used for packaging development tests per Method 516.3-5 to -10.

33.2 Specification describes time history. For many years, under MIL-STD-810, test motion was prescribed by its waveform (peak intensity, time duration and pulse shape). Any machine generating the desired pulse was OK; its authors wrongly ignored the frequency domain.

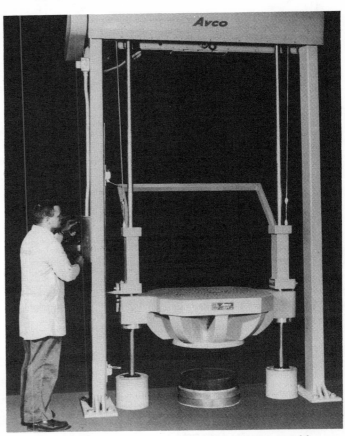

Fig. 33.9 Free-fall shock test machine. Dropping onto rubber impact cushion produces haversine pulse. Courtesy Avco.

33.2.1 Variety of "drop" machines for classical pulses. Motion arrest (Fig. 33.9) into rubber gives a haversine $\ddot{x}$ pulse. For sawtooth or square pulses, arrest with a soft lead casting; shape dictates the $\ddot{x}$ time history. Or (Fig. 33.10) use a fluid/orifice system to arrest motion. Some machines stretch elastic cord to add $\dot{x}$ for greater impact.

Classical pulses overtest items whose f_ns range 100-1,000 Hz and excite resonances not excited by field environments. Do not use them for "pyro" shock: too much low f_f energy, not enough high f_f energy. They are not oscillatory, as are field pyro shocks. The key to surviving classical pulses: ability to absorb KE.

We commonly include in the "drop" category the Avco machine of Fig. 33.11 and 33.12 (operates on shop compressed air.) The Hyge machine of Fig. 33.13 uses high pressure N_2 and hydraulics to vertically accelerate a carriage; brakes arrest motion.

33.2.2 Shaker for classical pulses. EH shakers offer long stroke for long-T pulses. A few EM shakers offer 50 mm (2 inch) D; most are 25 mm, 1 inch D. Long T, above say 0.003

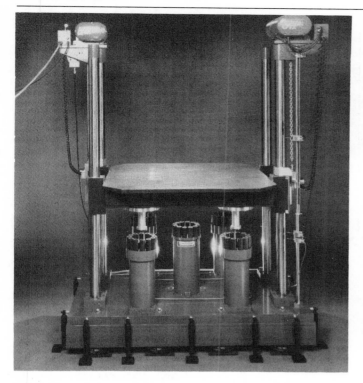

Fig. 33.10 Free-fall shock test machine with damage boundary programmers and seismic base. Courtesy Lansmont.

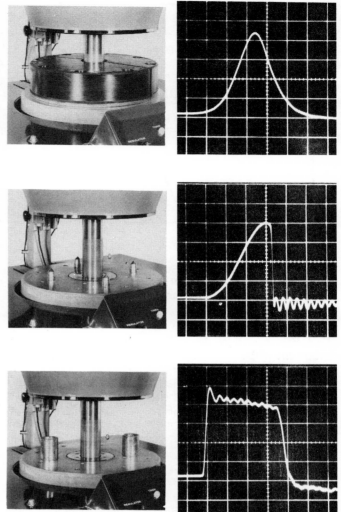

sec., requires much D. Shaker f_u limits restrict short T tests. Performing shock tests on an existing shaker saves buying a shock machine and extra test fixtures, also time transferring the load. Shock direction can be switch reversed.

Fig. 33.12 Three classical pulses (right) available with three stopping devices (left). Courtesy Avco.

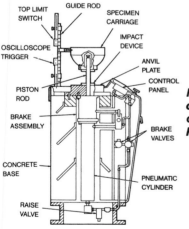

Fig. 33.11 Air pressure pulls carriage down against choice of stopping mechanisms (per Fig. 33.12). Courtesy Avco.

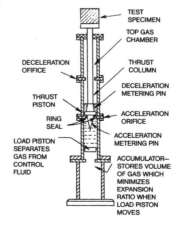

Fig. 33.13 Pneumatic cannon for several thousand g, few microsecond pulses. Courtesy Hyge.

33.2.3 Shaker for oscillatory shock. Classical pulse shapes do not occur in the real world. Better, tests typified by Fig. 32.1, brief intervals of multifrequency oscillation. "Drop" machines are useless. Use digital control per Sec. 33.4: EH shakers 0-200 Hz and EM shakers to say 2,000 Hz.

Service shocks often send transient wave motion through much intervening structure to equipment (and measurement) locations. On arrival, it is very complex. It briefly stimulates local structural (often lightly damped) and equipment resonances. While not ideal, shakers are much better for these tests than for classical pulse tests.

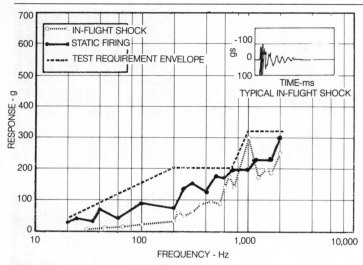

Fig. 33.14 Required Response Spectrum (RRS), upper, a special form of the SRS, somewhat more severe than the SRSs of two field events. A companion time history is also shown.

33.3 Specification describes the required response spectrum (RRS), a special form of the SRS (Sec. 32.6), in terms of its "damage potential." See Fig. 33.14. Such a broad-spectrum event excites many response modes to specified intensities. Neither the test pulse shape nor duration is specified. However, the RRS alone *could* be "cheated," met on a shaker by sweeping and varying a sine wave; responses would differ from field shock results. Thus the effective transient duration T_E (see Fig. 516.3-3 of MIL-STD-810D) is part of the specification. An "input" accelerometer signal is SRS analyzed to verify meeting the RRS.

33.3.1 Explosives. You might meet "pyro" RRS with explosives. Connect explosives to the test item via a fixture designed or "tuned" to emphasize the f_fs of greatest interest.

33.3.2 Shakers permit varying f_f content (of an electrical pulse into the PA). First attempts followed early analog equalization (Sec. 25.1) for random tests. Digital synthesis per Fig. 33.15 prevails today.

33.4 Digital shaker control (of sine and random vibration tests) was discussed in Sec. 15 and 26. We can digitally control shakers (most commonly EM) for shock tests. We quickly load shock software and change a front panel overlay.

33.4.1 Digital synthesis from RRS simplifies reproducing shock spectra. An early analog approach (pulse exciting a filter bank and summing the outputs) required much time and skill; accuracy and repeatability were poor.

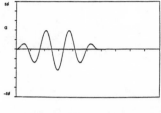
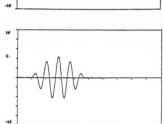

Fig. 33.16 Wavelets of 7 half cycles at f_f (upper) and 9 half cycles at $2f_f$ (lower); the latter has been delayed by 10 ms. Each is a sine wave weighted by a half sine wave of lower frequency. These are summed (with other wavelets) to produce the composite drive waveform.

Two techniques: Wavelet Equalization (WAE) and Transfer Function Equalization (TFE). Both use the same waveform to synthesize the shock spectrum, summing half sine wavelets. Each wavelet contains an odd number of half cycles weighted by a half sine wave per Fig. 33.16. However, the control processes are different.

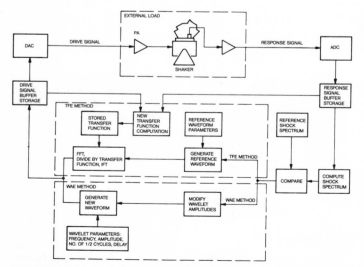

Fig. 33.15 Block diagram of a digital control system, modified for shock synthesis. Courtesy GenRad.

```
1 HEADING:WAE A TEST
2 SENSITIVITY(MV/G):10.
3 SHOCK RESP DEFN 0=ABS ACCEL 1=REL DISPL:0
4 DAMPING COEFF:.05
5 MAX FREQ:5000
6 # OF DECADES 0=2 1=2.3 2=2.6:0
7 WAVE PARAMETERS
        FREQ      AMPL    # OF 1/2 CYC   DELAY(MS)
   1    56.22     40.        5.            0.
   2    79.42     40.        5.            0.
   3    112.1     100.       9.            0.
   4    158.4     100.       9.            0.
   5    223.9     100.       9.            0.
   6    316.3     200.       9.            0.
   7    446.7     300.       9.            0.
   8    630.9     500.       9.            0.
   9    891.3     700.       9.            0.
  10   1259.     1000.       9.            0.
  11   1778.     1000.       9.            0.
  12   2511.     1000.       9.            0.
  13   3546.     1000.       9.            0.
  14   5013.     1000.       9.            0.
8 PEAK WAVELET AMPL(V):2.
9 AUTO MODE LEVEL SEQ 0=FULL 1=1/2 2=1/4 3=1/8 4=1/16 5=DONE
    FIRST:1
    NEXT:0
    NEXT:5
10 EXTERNAL TRIGGER MODE 0=NO 1=YES:0
11 ALARM BAND 1
    +DB LIMIT:3.
    -DB LIMIT:-6.
    UPPER FREQ, HZ:5000.
```

Fig. 33.17 Typical computer/operator dialog for a WAE RRS test.

Let's describe WAE set-up and testing. We first keyboard BW, sensitivity, damping, amplitude and other test parameters. At lines 1 and 2, Fig. 33.17, we key the test heading and accelerometer sensitivity. At line 3 we select SRS analysis computation based on test mass absolute $\ddot{x}$ or relative x of mass in an SDoF system. Enter C/C_c, f_u and number of decades at lines 4, 5 and 6. Line 7 calls for various wave parameters. We select the number of half cycles and suitable delay of each f_f. At line 8 we limit (here 2 volts) each wavelet's amplitude. At line 9 we program the sequence of test levels if using "automatic." Or we can manually select shock intensity and number of shocks. We decide (line 10) whether to trigger from an external source. At line 11 we select alarm (tolerance) limits. Now all parameters are stored on disk, ready for summing the individual wavelets into a composite waveform.

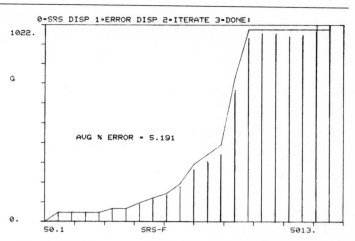

0=SRS DISP 1=ERROR DISP 2=ITERATE 3=DONE:

AUG % ERROR = 5.191

Fig. 33.20 SRS of the table motion per Fig. 33.19.

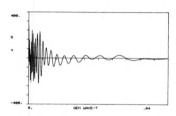

Fig. 33.18 Composite waveform from control to power amplifier.

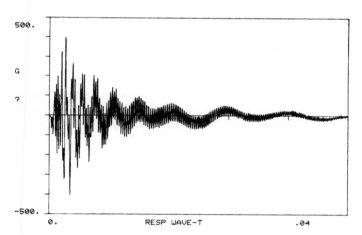

Fig. 33.19 Typical motion of loaded shaker to input waveform of Fig. 33.18.

The system sends (at say 1/16 full intensity) the summed waveform to the PA and shaker; the first attempt assumes a flat transfer function. The response waveform is sampled, analyzed, and compared to the RRS. The computer corrects any difference by adjusting wavelet amplitudes, ready for the next attempt, at greater intensity. Figs. 33.18, 33.19 and 33.20 illustrate an arbitrary but typical WAE input to the computer and the results. Data can be displayed and hard copied with 1/3, 1/6 or 1/12 octave resolution.

By contrast, the TFE method computes a transfer function and then generates a specific waveform to give the desired

RRS at the shaker. The internal steps: 1) CAL pulse goes to shaker; response is analyzed. 2) Transfer function is computed f_l to f_u; resolution of 4 or 8 Hz depends on f_u. 3) The computer generates a trial waveform. 4) The computer generates a compensated waveform and sends this to the shaker. 5) The SRS is computed and displayed. 6) The computer asks three questions: (a) is a new transfer function required? (b) is the transfer function OK? (c) terminate test? 7) If the operator chooses (a), the computer uses the last response from the shaker, step 5), and repeats steps 2) through 6). 8) When the operator is satisfied, he can continue testing at selected levels or terminate the test.

The required TFE drive signal is formed by first dividing the RRS by the transfer function, then generating an IFT to form the compensated waveform. This compensation is applied over the full test bandwidth.

WAE adjusts wavelet amplitude to achieve the desired spectrum. Whereas TFE produces a specific time history, WAE does not and makes no attempt to control every peak and notch in the spectrum. WAE only guarantees to reproduce the specified shock spectrum. WAE control is easier due to wider SDoF filter bandwidth. Because of its high resolution, TFE may saturate the PA or exceed shaker limits.

33.4.2 Digital synthesis of classical pulses per Fig. 33.7 utilizes high speed transfer function measurements. Lengthy calibrations and poor repeatability of analog shaker shock are avoided. Limitations in magnitude due to EM or EH shaker force, velocity and displacement limits still apply.

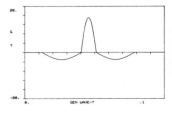

Fig. 33.21 Older classical half-sine pulse, obtainable with digital shaker control. Negative "tail" area = positive pulse area.

To reproduce classical shock pulses on EM shakers, final $\dot{x}$ and x must be zero. This requires adding negative operator-selected pre- and post-pulse half sine pulses or "tails" per Fig. 33.21.

Operation resembles Sec. 33.4.1. The operator loads necessary software and keys required parameters into memory per Fig. 33.22. Note several available pulse shapes: half sine (HS), square wave (SQ), sawtooth (ST), triangular (TR), double pulse (DP), as well as arbitrary analog or digital waveforms per Sec. 33.4.3. At line 9 the operator selects calibration pulse amplitude, here 1 volt. Note (from lines 11 and 12) that

```
1 HEADING:TEST A SHOCK
2 SENSITIVITY(MV/G):10.
3 WAVE TYPE: 0=HS 1=SQ 2=ST 3=TR 4=DP 5=ANALOG 6=DIGITAL:0
4 DURATION(MS):11.
5 AMPLITUDE(G):15.
6 % IN TAILS:10.
9 CAL PULSE AMPL(V):1.
10 AUTO MODE LEVEL SEQ 0=FULL 1=1/2 2=1/4 3=1/8 4=1/16 5=DONE
   FIRST:2
   NEXT:1
   NEXT:0
   NEXT:5
11 SHOCK RESP DEFN 0=ABS ACCEL 1=REL DISPL:0
12 DAMPING COEFF:.05

CORRECTIONS 0=NO 1=YES:
```

Fig. 33.22 Typical operator/computer dialog for, in this case, an older half-sine pulse. Operator responses are underlined.

an SRS can be calculated to compute absolute $\ddot{x}$ or relative x with the defined damping coefficient. When he has entered all parameters, the operator tells the computer to generate the pulse to be processed. At line 10 he selects the sequence of levels in the AUTO mode. At line 11 he selects a particular kind of SRS. At line 12 he selects a damping coefficient.

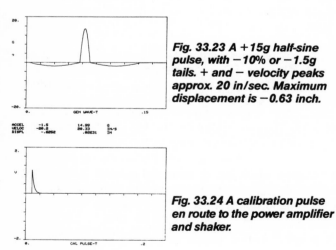

Fig. 33.23 A +15g half-sine pulse, with −10% or −1.5g tails. + and − velocity peaks approx. 20 in/sec. Maximum displacement is −0.63 inch.

Fig. 33.24 A calibration pulse en route to the power amplifier and shaker.

The system is now ready. The operator sends the calibration pulse of Fig. 33.24 to the shaker. The response is analyzed and a transfer function is calculated, resulting in the compensated waveform, Fig. 33.25. He can output a compensated wave several times at increasing pre-test intensities. This calibrates the system for the desired pulse, specifying a new transfer function to be computed at each step. Finally, he applies one or more full test pulses or terminates testing. An

SRS or FFT of each shock pulse can be displayed and a hard copy made. A switch can reverse the system's drive polarity to reverse the shock.

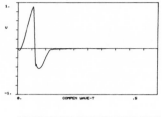

Fig. 33.25 Compensated waveform, an attempt to achieve an arbitrary half sine pulse.

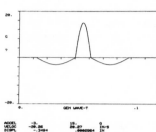

Fig. 33.26 The same +15g 11 ms half-sine pulse, now with -20% or -3g tails. + and -velocity peaks approx. 20 in/sec. Maximum displacement is now -0.35 inch.

A caution re EM shakers for classical pulse shock testing: be aware of X, the maximum armature motion. The X of Fig. 33.23 was 0.625 inch; with a shaker X limit of 0.5 inch (D = 1 inch), considerable "bias" was needed to avoid hitting the stops. Increased tail amplitude per Fig. 33.26 increases the available X. However, you should consider tolerances on pulse shape, as in Fig. 33.27.

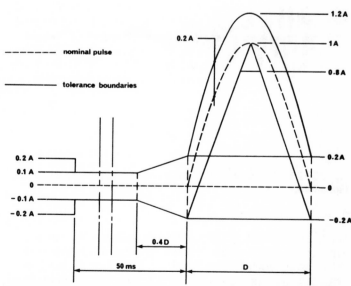

Fig. 33.27 Tolerances sometimes found on specified older half-sine pulses.

Suppliers have recently optimized pre- and post-load pulses. Fig. 33.28 shows two different methods. The resultant X is significantly less than formerly. This permits higher A for a given X. Also lower PA requirements.

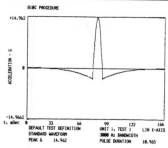

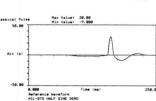

Fig. 33.28 Optimized half-sine pulses. Courtesy Scientific-Atlanta (upper) and courtesy GenRad (lower).

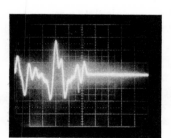

Fig. 33.29 Arbitrary tape recorded 64 ms duration oscillatory pulse used in test of GenRad computer capability.

33.4.3 Digital reproduction of oscillatory pulses,
seems closer to the "real world." Envision testing with the duplicate of an arbitrary field $\ddot{x}$ pulse per Fig. 33.29. First we loaded software and responded to several questions per Fig. 33.30. At line 3 we chose *5*, as we had an analog signal on tape. (We might have stored it digitally on disk.) Line 7 (lines 4, 5 and 6 are omitted for this pulse type) asks for input buffer duration (32 ms to 32 sec.) We chose 64 ms. After a few more questions, the system was ready to store the waveform.

```
ENTER PARAMETERS 0=NO 1=YES:1
INPUT 0=RDR 1=KYBD 2=DISK:1

1  HEADING:SHK TEST
2  SENSITIVITY(MV/G):100
3  WAVE TYPE: 0=HS 1=SQ 2=ST 3=TR 4=DP 5=ANALOG 6=DIGITAL:5
7  BUFFER DURATION(MS):64
   BUFFER DURATION(MS):64
8  # OF G PER VOLT:20
9  CAL PULSE AMPL(V):1
10 AUTO MODE LEVEL SEQ 0=FULL 1=1/2 2=1/4 3=1/8 4=1/16 5=DONE
   FIRST:3
   NEXT:1
   NEXT:0
   NEXT:5
11 SHOCK RESP DEFN 0=ABS ACCEL 1=REL DISPL:0
12 DAMPING COEFF:.05

CORRECTIONS 0=NO 1=YES:0
LIST 0=NO 1=YES:0
SAVE 0=NO 1=YES:0
```

Fig. 33.30 System/operator dialog — slightly different from Fig. 33.22. Operator responses are underlined.

We adjusted the system to capture the wave (Fig. 33.31). We checked the calculated $\ddot{x}$, $\dot{x}$ and x against our shaker capabilities. Next we outputted a calibration pulse per line 9

for calculating an inverse transfer function from which the computer generated a compensated waveform (Fig. 33.32) to drive the shaker. Fig. 33.33 shows the final shaker pulse, close to Fig. 33.29.

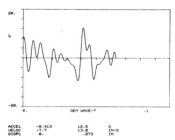

Fig. 33.31 Desired waveform, as stored in memory.

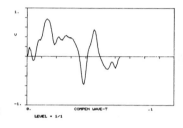

Fig. 33.32 Compensated waveform, generated by computer.

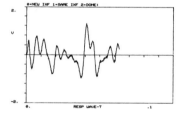

Fig. 33.33 Shaker motion, a fair reproduction of Fig. 33.29.

33.4.4 Shaker force ratings
(for shock tests) are greater than for continuous sine (Sec. 12.3) or continuous random (Sec. 23.10). "What A could you achieve with a huge PA?" is a difficult question, partly depending upon the test load. At low vibration f_fs, say 10-200 Hz, NASA has about doubled the F = MA rating without damage. Shaker electrical resonance (Sec. 14.4.1) probably helped. PAs no doubt delivered high current peaks. Over perhaps 200-4,000 Hz you might use a multiple of 10. Big shakers can achieve 300g peaks, with Q = 10 SRS approaching 1,000g at higher f_fs. With smaller shakers, you might get SRS to 2,000g.

33.5 Miscellaneous shock machines
exist for special purposes. Transportation tests determine if cargo is suitably packaged to survive transport. "Conbur" and "rotating drum" tests are largely obsolete. Sec. 10.4 mentioned mechanical shakers. Consider (Sec. 11.5) EH shakers for cargo tests; large D and large F for low f_f random vibration are their main advantages.

33.5.1 Drop tests.
Containers are dropped onto a concrete floor or other unyielding surface, onto their flat surfaces,

edges and corners. Devices suggested by Fig. 33.34 hold, then drop containers from specified heights.

Fig. 33.34 Support arm pulls away, allowing container to free fall, impacting floor. More realistic than package and carriage stopping together. Courtesy MTI/LAB.

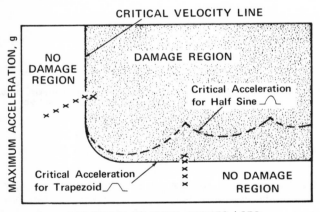

CRITICAL VELOCITY LINE

MAXIMUM ACCELERATION, g

NO DAMAGE REGION

DAMAGE REGION

Critical Acceleration for Half Sine

Critical Acceleration for Trapezoid

NO DAMAGE REGION

VELOCITY CHANGE, INCHES / SEC.
(Proportional to Design Drop Height)

Fig. 33.35 Plot of damage boundary. Courtesy Lansmont.

33.5.2 Packaging development tests are performed on shock test machines (see Sec. 33.2.1) whose peak A and whose change in velocity or ΔV are independently variable. Mount an unpackaged (hardmounted) product on the machine, set to produce a half sine pulse of low ΔV. Check the product for "damage." Repeat, each time increasing ΔV, until "damage" occurs. Mark the ΔV values per Figure 33.35, with the critical ΔV line through the next-to-final point (more conservative than final point). Repair or replace the product.

Repeat the sequence with trapezoidal or square pulses, with test ΔV at least 1.57 $\times$ critical ΔV, by dropping the carriage further. Keep test ΔV constant by varying acceleration and duration. Mark the g values per Fig. 33.35. Eventually "damage" will occur; draw the critical $\ddot{x}$ line through the next-to-final point, per Fig. 33.35.

Now that the hardmounted product has been tested, equate critical ΔV to possible drop heights by $\Delta V = \sqrt{2gh}$. If a small product might be dropped 30 inches, calculate its potential ΔV as 152 in/sec or 3.86 m/s. If the item's critical ΔV exceeds this, no cushioning is needed.

Your package designer will select appropriate packaging and cushioning to sufficiently mitigate 30-inch "real world" drops so your critical g is not exceeded. (Cushioning also provides isolation against vibration, per Sec. 1.1.5.) See A.S.T.M. procedures D3331-74T and D3332-74T, also "Product Fragility/Assessment," by W. I. Kipp of Lansmont. Also Federal Standard 110 and MIL-P-116. The word "damage" in Sec. 33.5.2 can be interpreted variously. It might refer to a scratch, to a lever popping off or to real interference with functioning.

33.5.3 Vehicular crash tests. Several machine types deal with automotive crash safety. One uses gas pressure in a horizontal ram (much like the "Hyge" unit of Fig. 33.13) to throw a passenger car backward. Instrumented mannequins inside react as though the vehicle had stopped abruptly. Variations slam a test mass into an automobile's front or rear or side. Sometimes a bumper is under test, rather than passenger restraints. Another variation tests poles (to be located along highways); the poles are *supposed* to fail.

We greatly appreciate the assistance of Glenn Wasz of TRW, Jon Wilson of Endevco and Randall R. Eager of Lansmont in preparing Sec. 33.

Section 34
Abbreviations and Definitions

34.1 **Abbreviations**
34.2 **Definitions**

34.1 Abbreviations used in this text:

A	Acceleration, zero-to-peak or RMS value
a or $\ddot{x}$	Acceleration, instantaneous
AA	Arithmetic average
ac	Alternating current
ADC	A/D or analog to digital converter
AGC	Automatic gain control (electronic circuit)
AM	Amplitude modulation
ANSI	American National Standards Institute
APD	Amplitude probability density
ASD	Acceleration spectral density (also PSD)
ASME	American Society of Mechanical Engineers
ASQC	American Society for Quality Control
ASTM	American Society for Testing and Materials
B or BW	Bandwidth (as in filters, or in spectra)
C	Amount of damping present
C_c	Amount of damping called critical
CCTV	Closed circuit television
CERT	Combined environment reliability test
cg	Center of gravity
cpm	Cycles per minute
cps	Cycles per second, now called hertz or Hz
CRT	Cathode ray tube (or terminal)
D	Displacement, peak-to-peak or RMS value
DAC	D/A or digital to analog converter
dc	Direct current
DoF	Degrees of freedom
E	Voltage
E_s	Voltage from a standard transducer
E_u	Voltage from an unknown transducer
EH	Electrohydraulic (as in shakers)
EHPV	Electrohydraulic pneumatic valve
EM	Electromagnetic (as in shakers)
EMPV	Electromagnetic pneumatic valve
EVM	Engine vibration monitoring (as on aircraft)
F	Force, zero-to-peak or RMS value
FFT	Fast Fourier transform
f_c	Crossover frequency (sine testing)
f_f	Forcing frequency
f_l	Lower bounding frequency of a spectrum or a sweep
f_n	Natural frequency of a structure, or frequency of a notch
f_o	Center frequency of analyzer filter
f_p	Frequency of a peak
f_s	Natural frequency of a sensor (as mounted)
f_T	Total of several natural frequencies (combined)
f_u	Upper bounding frequency of a spectrum or a sweep
FM	Frequency modulation
FSP	Floating shock platform
GenRad	General Radio Co. (see 35.1)

H-P	Hewlett-Packard (see 35.1)
Hz	Frequency in hertz or cycles per second
I	Current
IC	Integrated circuit
ID	Identification
IFT	Inverse Fourier transform
in	Inches
IR	Infrared
IRD	International Research & Development Corp. (see 35.1)
ISA	Instrument Society of America
IES	Institute of Environmental Sciences
lb	Pounds (weight, sometimes force)
lbf	Pounds force
K	Arbitrary symbol for an unchanging quantity
K	Spring stiffness
KE	Kinetic energy
kva	Kilovolt-amperes
kw	Kilowatts
L	Length
LDS	Ling Dynamic Systems (see 35.1)
LVDT	Linear variable differential transformer
LWS	Light weight shock (U.S. Navy) test machine
m	Distance in metres
M	Mass
MDoF	Multiple degree of freedom
mph	Miles per hour
MBIS	MB Instrument Systems (see 35.1)
MUX	Multiplexer
MWS	Medium weight shock (U.S. Navy) test machine
N	Newtons-force
NBS	The U.S. National Bureau of Standards
NoF	Number of filters
NRL	U.S. Naval Research Laboratory, Washington, D.C.
NTIS	National Technical Information Service, Washington, D.C.
P	Preload, holding force, as in a bolt under tension
P_I	Preload, initial, immediately after tightening
P_R	Preload, residual, after time has passed
PA	Power amplifier
Pa	Pascals—unit of pressure
PCM	Pulse code modulation
pc	Picocoulomb, 10^{-12} coulomb of electrical charge
PE	Piezoelectric, as in accelerometers
pf	Picofarad, 10^{-12} farad of electrical capacitance
PR	Piezoresistive, as in accelerometers
PSD	Power spectral density (also ASD)
Q	Resonant magnification factor (structure or filter)
RF	Radio frequency
RFP	Request for proposal
RFQ	Request for quotation

RMS	Root-mean-square	TF	Tracking filter
RRS	Required response spectrum	TIT	Tustin Institute of Technology
RTA	Real time analyzer	Tr	Transmissibility
RVDT	Rotary variable differential transformer	TRMS	True root mean square
s	Time in seconds	TRS	Test response spectrum
SDoF	Single degree of freedom	TVDT	Torsional variable differential transformer
SI	Systeme Internationale or "metric" system	U-D	Unholtz-Dickie (see 35.1)
S/N	Signal-to-noise ratio, as in electrical measurements	V	Velocity, zero-to peak or RMS value
		v or $\dot{x}$	Velocity, instantaneous
SVIC	Shock & Vibration Information Center (see 35.3.2)	VCO	Voltage controlled oscillator
SRS	Shock response spectrum	W	Weight
S_s	Sensitivity of a standard sensor	X	Displacement, zero-to-peak or RMS value
S_u	Sensitivity of an unknown sensor	x	Displacement, instantaneous
T	Time interval, as period of one cycle or voltage decay	$\ddot{x}$	Acceleration, instantaneous
T_E	Effective transient duration (of a shock pulse)		

34.2 Definitions of some terms used in vibration and shock.

Acceleration. Acceleration is rate of change of velocity with time (denoted as dv/dt or d^2x/dt^2 or $\ddot{x}$), usually along a specified axis, usually expressed in *g* or gravitational units. It may refer to angular motion.

Accelerometer. A device (a *Sensor* or *Transducer* or *Pickup*) for converting $\ddot{x}$ to an electrical signal.

Accuracy. The capability of an instrument to indicate the true value. Do not confuse with *inaccuracy* (sum of hysteresis + non-linearity + drift + temperature effect, etc.) nor with *Repeatability*.

Amplitude. The magnitude of variation (in a changing quantity) from its zero value. Always modify it with an adjective such as *peak*, *RMS*, *average*, etc. May refer to displacement, velocity, acceleration, force or pressure.

Angular Frequency. Angular frequency (also known as circular frequency ω is the torsional vibration frequency in radians per second. Or multiply by 2π and express in cycles per second (cps) or hertz (Hz).

Average. Refer to a textbook on electrical engineering. In the exclusive case of a sine wave, the average value is 0.636 × peak value.

Balancing. (mechanical) Adjusting mass distribution in a rotating element, to reduce vibratory forces generated by rotation.

Broadband. Vibration signals which are unfiltered. Signals at all frequencies contribute to the measured value.

Calibration. (As applied to vibration sensors) an orderly procedure for determining sensitivity as a function of frequency, temperature, amplitude, etc. Yields deviations from correct values used for inferring true magnitudes from indicated magnitudes. Calibration may also refer to adjusting an instrument, lessening deviations from a standard sensitivity.

Compliance. The reciprocal of *Stiffness*, i.e. displacement ÷ force.

Critical Frequency. A particular resonant frequency (see Resonance) at which damage to (or degradation in performance) is likely.

Cycle. The complete sequence of instantaneous values of a periodic event, during one period.

Damping. Dissipation of oscillatory or vibratory energy with motion or with time. *Critical* damping is that value that provides most rapid response to a step function without overshoot. $C_c = 2\sqrt{KM}$. Damping ratio is then C/C_c.

Decade. The interval between two frequencies which differ by exactly 10:1.

DeciBel. Ratios of identical quantities are expressed in decibel or deciBel or dB units. Magnitude thus refers to some standard value, in terms of the base 10 logarithm of that ratio. In measuring acoustic or vibration *power* (as in PSD or ASD of random vibration), the number of dB = $10 \log_{10} P/P_0$. P_0, the reference level, equals 0 dB. In measuring the more common *voltage*-like quantities such as $\ddot{x}$, the number of dB = $20 \log_{10} E/E_0$. E_0, the reference level, equals 0 dB.

Degrees of freedom. In mechanics, the total number of directions of motion (of all the points being considered) on a structure being modeled or otherwise evaluated. In statistics, the number of independent variables used in constructing a mathematical model representing some collection of random variables.

Deterministic vibration. A vibration whose instantaneous value at any future time can be predicted by an exact mathematical expression. Sinusoidal vibration is the classic example. Complex vibration is less simple (two or more sinusoids).

Displacement specifies change of position, usually measured from mean position or position of rest. Usually applies to uniaxial, less often to angular motion.

Distortion. In mechanics, any unwanted motion. If sinusoidal motion were desired at a fundamental frequency, any motion at harmonics or subharmonics of that frequency, or any mechanical "hash" (perhaps due to parts colliding). In electronic measurements, any unwanted signal; e.g. amplifiers may generate unwanted signals.

Duration of a shock pulse is how long it lasts. For "classical" pulses, time is usually measured between instants when the amplitude is greater that 10% of the peak value.

Filter. An electronic device to pass certain frequencies (pass band) but block other frequencies (stop band). Classified as *low-pass* (high-stop), *high-pass* (low-stop), *band-pass* or *band-stop*.

Forced vibration. The vibratory motion of a system caused by some mechanical excitation. If excitation is periodic and continuous, motion eventually becomes steady-state.

Free vibration. Free vibration occurs without forcing, as after a reed is plucked.

Frequency. The reciprocal of the period in seconds (of a periodic function) (1/T). Usually given in hertz (Hz), meaning cycles per second (cps).

Frequency response. The portion of the frequency spectrum which a device can cover within specified limits of amplitude error.

Fundamental frequency. The number of cycles per second of the lowest-frequency component of a complex, cyclic motion. (See also *Harmonic* and *Subharmonic*.)

Frequency spectrum. A description of the resolution into frequency components, giving the amplitude (sometimes also phase) of each component.

Fundamental mode of vibration. That mode having the lowest f_n.

g. The acceleration produced by Earth's gravity. By international agreement, the value for 1 gravitational unit is 9.80665 m/s² = 386.087 in/sec² = 32.1739 ft/sec².

g units or **gravitational units.** A way to express $\ddot{x}$, in terms of the gravitational constant, is equal to $\ddot{x}$ in/sec² ÷ 386.087 in/sec² or to $\ddot{x}$ m/s² ÷ 9.80665 m/s².

Harmonic. A sinusoidal quantity having a frequency that is an integral multiple (×2, ×3, etc.) of a fundamental (×1) frequency.

Hash. Distortion on a signal viewed on an oscilloscope trace (slang).

Impact. A collision between masses.

Impulse. The integral of force over a time interval.

Induced environments. Conditions generated by operating an equipment, as opposed to natural environments.

Inertance (or accelerance.) The ratio of acceleration to force.

Input. The mechanical motion, force or energy applied to a mechanical system, e.g. the vibratory input from shaker to test item. Or an electrical signal, e.g. from an oscillator to the PA driving a shaker.

Input control signal. Originates in a control sensor; some-

times selected between or averaged between several sensors. Used to regulate shaker intensity. (May originate in a force sensor for force-controlled testing.)

Intensity. The severity of a vibration or shock. Nearly the same meaning as *Amplitude,* defined earlier, but less precise, lacking units.

Isolation. A reduction in motion severity, usually by a resilient support. A shock mount or isolator attenuates shock. A vibration mount or isolator attenuates steady-state vibration.

Jerk. The rate of change of acceleration with time.

Linear system. A system is linear if its response is directly proportional to excitation, for every part of the system.

Mass. A physical property, dynamically computed as acceleration ÷ force. Statically computed as W (which can be measured on a butcher scale) ÷ g. Ordinary structures are not pure masses as they contain reactive elements, i.e. springs and damping.

Mean. A value intermediate between quantities under consideration. The mean $\ddot{x}$ on a shaker must be zero — no steady-state acceleration. But a vehicle can have steady-state $\ddot{x}$.

Mechanical impedance. The ratio of force to velocity, where the velocity is a result of that force only. Its reciprocal, *mobility*, is today more favored.

Mode. A characteristic pattern in a vibrating system. All points reach their maximum x at the same instant.

Natural environments. Conditions occurring in nature; effects are observed whether an equipment is at rest or in operation.

Natural frequency. f_n, the frequency of of an undamped system's free vibration; also, the frequency of any of the normal modes of vibration. f_n drops when damping is present.

Noise. The total of all interferences in a measurement system, independent of the presence of signal.

Notch. Minimum spectral value, at f_n. Also, the deliberate reducing of a test spectrum at certain f_s.

Octave. The interval between two frequencies differing by exactly 2:1.

Oscillation. Variation with time of a quantity such as F, stress, pressure, x, v, a or jerk. Usually implies some regularity (as in sinusoidal or complex vibration).

Peak. Extreme value of a varying quantity, measured from the zero or mean value. Also, a maximum spectral value, occuring at f_p.

Peak-to-peak value. The algebraic difference between extreme values (as D = 2X).

Period. The smallest interval of time in which a cyclic vibration repeats itself.

Periodic vibration. (See also *Deterministic vibration*.) An oscillation whose waveform regularly repeats. Compare with *Probabilistic vibration*.

Phase. (Of a periodic quantity), the fractional part of a period between a reference time (such as when x = 0) and a particular time of interest; or between two motions or electrical signals having the same fundamental frequency.

Pickup. See *Transducer*.

Piezoelectric transducer. One which depends upon deformation of its sensitive crystal or ceramic element to generate electrical charge and voltage. Many present-day accelerometers are PE.

Piezoresistive transducer. One which depends upon deformation of its sensitive resistive element (certain semiconductors — greater resistance change than wire for a given deformation). (See *Strain-gage transducer*.)

Platform Per MIL-STD-810D, any vehicle, surface or medium that carries an equipment. For example, an aircraft is the carrying platform for internally installed avionics equipment and externally mounted stores. The land is the platform for a ground radar set, and a man for a hand carried radio.

Power spectral density or PSD. Describes the power of random vibration intensity, in mean-square $\ddot{x}$ per frequency unit, as g^2/Hz or m^2/s^3. Acceleration spectral density or ASD is preferred abroad.

Precision. The smallest distinguishable increment (almost the same meaning as *Resolution*); deals with a measurement system's possible or design performance.

Probabilistic vibration. (As compared to *Deterministic vibration*), one whose magnitude at any future time can only be predicted on a statistical basis.

Quadrature motion. (Or Side or Lateral motion or Cross-talk), any motion perpendicular to the reference axis. Shakers are supposed to have zero quadrature motion.

Quadrature sensitivity. (Or Side or Lateral or Cross-Talk sensitivity) of a vibration sensor is its sensitivity to motion perpendicular to sensor principal axis. Commonly expressed in % of principal axis sensitivity.

Random vibration. (See *Probabilistic vibration*.) One whose instantaneous magnitudes cannot be predicted. Adjective "Gaussian" applies if they follow the Gaussian distribution. May be *broad-band*, covering a wide, continuous f_f range, or *narrow-band*, covering a relatively narrow f_f range. No periodic or deterministic components.

Range. A statement of the upper and lower limits over which an instrument works satisfactorily.

Repeatability. (1) The maximum deviation from the mean of corresponding data points taken under identical conditions. (2) the maximum difference in output for identically-repeated stimuli (no change in other test conditions). Do not confuse with *Accuracy*.

Response. The vibratory motion or force that results from some mechanical input.

Response signal. The signal from a "response sensor" measuring the mechanical response of a mechanical system to and input vibration or shock.

Resolution. The smallest change in input that will produce a detectable change in an instrument's output. Differs from *precision* in that human capabilities are involved.

Resonance. Forced vibration of a true SDoF system causes resonance when $f_f = f_n$, when any f_f change decreases system response. (See also *Critical frequency*.) Therefore resonance represents maximum sprung mass response, if f_f is varied while input F is held constant.

Ringing. Continued oscillation after an external force or excitation is removed, as after a guitar string is plucked.

RMS or **Root-Mean-Square** value. The square root of the time-averaged squares of a series of measurements. Refer to a textbook on electrical engineering. In the exclusive case of a sine wave, the RMS value is 0.707 × the peak value.

Sensitivity. Of a mechanical-to-electrical sensor or pickup, the ratio between electrical signal (output) and mechanical quantity (input).

Sensor. (See *Transducer*.)

Self-induced vibration. Also called self-excited vibration, results from conversion of non-oscillatory energy into vibration, as wind exciting telephone wires into mechanical vibration.

Shock machine. Or shock test machine, a device for subjecting a system to controlled and reproducible mechanical shock pulses.

Shock pulse. A transmission of kinetic energy into a system in a relatively short interval compared with the system's natural period. A natural decay of oscillatory motion follows. Usually displayed as time history, as on an oscilloscope.

Shock response spectrum (or SRS), a plot of maximum response of SDoF systems vs. their f_ns, responding to an applied shock.

Signal conditioner. An amplifier following a sensor, prepares the signal for succeeding amplifiers, transmitters, readout instruments, etc. May also supply sensor power.

Simple harmonic motion. Periodic vibration that is a sinusoidal function of time.

Slew rate. The maximum rate at which an instrument's output can be changed by some stated amount.

Standard deviation σ. A statistical term: the square root of the variance, i.e., the square root of the mean of the squares of the deviations from the mean value.

Stationarity. A property of probabilistic vibration if the PSD (or ASD) and the probability distribution are constant.

Steady state vibration. Periodic vibration for which the statistical measurement properties (such as the peak, average, RMS and mean values) are constant.

Stiffness. The ratio of force (or torque) to deflection of a spring-like element, as $K = W/\delta$.

Strain-gage transducer. A changing-resistance sensor whose signal depends upon sensitive element deformation. In an unbonded wire strain-gage accelerometer, inertia affects a mass supported by wires; they change resistance proportional to $\ddot{x}$. Term may include piezoresistive accelerometers.

Stress Screening. A modern electronics production tool for precipitating latent defects such as poorly-soldered connections. Utilizes random vibration + temperature shock.

Subharmonic. A sinusoidal quantity having a frequency that is an integral submultiple ($\times 1/2$, $\times 1/3$, etc.) of a fundamental ($\times 1$) frequency.

Tailoring. Selecting or altering test procedures, conditions, values, tolerances, measures of failure, etc., to simulate or exaggerate the environmental effects of one or more forcing functions.

Tracking filter. A narrow bandpass filter whose f_o follows an external synchronizing signal.

Time constant. The interval needed for an instrument's output to move 63% of its ultimate shift as a result of a step change input.

Transducer (or *Pickup* or *Sensor*). A device which converts some mechanical quantity into an electrical signal.

Transient vibration. Short-term vibration of a mechanical system.

Transmissibility. In steady-state vibration, Tr is the non-dimensional ratio of response motion/input motion: two displacements, two velocities or two accelerations. The maximum Tr value is the mechanical "Q" of a system. At resonance, Tr is maximum.

Velocity. Rate of change of x with time (usually denoted as dx/dt or $\dot{x}$) usually along a specified axis; it may refer to angular motion as well as to uniaxial motion.

Vibration. Mechanical oscillation or motion about a reference point of equilibrium.

Vibration machine (or Exciter or Shaker). One which produces controlled and reproducible mechanical vibration for vibration testing mechanical systems, components, and structures.

Vibration meter. An apparatus (usually an electronic amplifier, detector and readout meter) for measuring electrical signals from vibration sensors. May read displacement, velocity and/or acceleration.

Weight. That property of an object that can be weighed, as on a scale; gravitational force on an object.

White random vibration. That broad-band random vibration in which the PSD (ASD) is constant over a broad frequency range.

Section 35
Sources of Vibration and Shock Information

35.1 Manufacturers.
35.1.1 Shaker systems

ACG, Inc.
232 Front Ave.
West Haven, CT 06516
203/776-6990

ACOUSTIC POWER SYSTEMS, INC.
2151-G Las Palmas Dr.
Carlsbad, CA 92008
619/438-4848

ALPHA-MONARCH
3009 Wildflower Drive
Dallas, TX 75229
214/620-0021

BRANFORD VIBRATOR CO.
Box 427
New Britain, CT 06051
203/224-3183

BRUEL & KJAER — see Sec. 35.1.2.

DERRITRON ELECTRONICS, Ltd.
Sedlescombe Road N.
Hastings, E. Sussex TN3 3XB
England 0424/754321

ENVIRONMENTAL SCREENING
   TECHNOLOGY, INC.
285 Kollen Park Drive
Holland, MI 49423
616/396-1485

INSTRON ENVIRONMENTAL, LTD.
Coronation Road
High Wycombe, Bucks
HP12 3SY ENGLAND
0494/33333

INSTRON, INC.
100 Royall St.
Canton, MA 02021
617/828-2500

LANSMONT
Box 1390
Monterey, CA 93942
408/373-4791

LING DYNAMIC SYSTEMS, LTD.
Baldock Road
Royston, Herts SG8 5BQ
England 0763/42424

60 Church St.
Yalesville, CT 06492
203/265-7966

LING ELECTRONICS, INC.
   Subsidiary of MTI
4890 E. La Palma Avenue
Anaheim, CA 92806
714/774-2000 or 779-1900

L.A.B, INC.
   Subsidiary of MTI
968 Albany Shaker Rd.
Latham, NY 12110
518/785-2221

MBIS, INC.
25865 Richmond Road
Bedford Heights, OH 44146
216/292-5850

M/RAD
71 Pine Street
Woburn, MA 01801
617/935-5940

MTS SYSTEMS CORP.
Box 24012
Minneapolis, MN 55424
612/937-4000

SCREENING SYSTEMS, INC.
4 Faraday
Irvine, CA 92714
714/855-1751

SHINKEN
c/o Bokosui Brown Co., Ltd.
Daini-Marutaka Bldg.
13-8, Ginza 7-chome
Chuo-Ku, Tokyo 104, Japan
543/8831-8

TEAM CORP.
9949 Hayward Way
South El Monte, CA 91733
818/442-3240

THERMOTRON INDUSTRIES, INC.
291 Kollen Park Drive
Holland, MI 49423
616/392-1491

UNHOLTZ-DICKIE CORP.
6 Brookside Drive
Wallingford, CT 06492
203/265-3929

VIBRATION TEST SYSTEMS
10246 Clipper Cove
Aurora, OH 44202
216/562-5729

ZONIC CORP.
2000 Ford Circle
Milford, OH 45150
513/248-1911

35.1.2 Accelerometers

BRUEL & KJAER
Linde Allee
DK-2850 Naerum, Denmark
01/800500

185 Forest Street
Marlborough, MA 01752
617/481-7000

COLUMBIA RESEARCH
LABORATORIES, Inc.
MacDade Blvd. at Bullens Lane
Woodlyn, PA 19094
215/872-3900

DYTRAN INSTRUMENTS, INC.
8523H Canoga Ave.
Canoga Park, CA 91304
818/700-7818

ENDEVCO
30700 Rancho Viejo Road
San Juan Capistrano, CA 92675
714/493-8181

ENTRAN DEVICES
10 Washington Avenue
Fairfield, NJ 07006
201/227-1002

GULTON
1644 Whittier Avenue
Costa Mesa, CA 92627
714/642-2400

KISTLER INSTRUMENTS
75 John Glenn Drive
Amherst, NY 14120
716/691-5100

PCB PIEZOELECTRONICS
3425 Walden Avenue
Depew, NY 14043
716/684-0001

SETRA SYSTEMS
45 Nagog Park
Acton, MA 01720
617/263-1400

UNHOLTZ-DICKIE — see Sec. 35.1.1.

VIBRA-METRICS
385 Putnam Avenue
Hamden, CT 06517
203/288-6158

VIBRO-METER (formerly BBN)
50 Moulton St.
Cambridge, MA 02138
617/491-0091

WILCOXON RESEARCH
Box 5798
Bethesda, MD 20814
301/770-3790

35.1.3 Analyzers and controls

BENTLY-NEVADA
Box 157
Minden, NV 89423
702/782-3611

BRUEL & KJAER — see Sec. 35.1.2.

GEN RAD
2855 Bowers Avenue
Santa Clara, CA 95051
408/727-4400

HEWLETT-PACKARD
Box 69
Marysville, WA 98270
206/335-2258

HONEYWELL TEST
INSTRUMENTS DIV.
Box 5227
Denver, CO 80217
303/773-4700

IRD MECHANALYSIS, INC.
6150 Huntley Road
Columbus, OH 43229
614/885-5376

MBIS — see Sec. 35.1.1.

METRIX INSTRUMENT CO.
1711 Townhurst Drive
Houston, TX 77043
713/461-2131

ONO SOKKI
c/o SHIGMA
80 Martin Lane
Elk Grove Village, IL 60007
312/640-8640 or
800/323-0315

PALOMAR TECHNOLOGY
 INTERNATIONAL
Box 1966
Carlsbad, CA 92008
714/496-0873

PMC/BETA CORP.
4 Tech Circle
Natick, MA 01760
617/237-6920

RION CO., Ltd.
Ikeda Bldg.
7-7, Yoyogi 2-chome
Shibuya-ku, Tokyo 151, Japan
379-3251

SPECTRAL DYNAMICS/
SCIENTIFIC-ATLANTA
Box 23575
San Diego, CA 92123
619/268-7100

STRUCTURAL MEASUREMENT
SYSTEMS, Inc.
645 River Oaks Pkwy.
San Jose, CA 95134
408/263-2200

SYNERGISTIC TECHNOLOGY, Inc.
20065 Stevens Creek Blvd.
Cupertino, CA 95014
408/253-5800

TAPED RANDOM CORP.
Box 133
Teaneck, NJ 07666
201/836-7392

TRIG-TEK, INC.
423 S. Brookhurst Street
Anaheim, CA 92804
714/956-3593

UNHOLTZ-DICKIE — see Sec. 35.1.1.

WAVETEK/NICOLET/ROCKLAND
10 Volvo Drive
Rockleigh, NJ 07647
201/767-7900

35.1.4 Fixtures

AVCO ELECTRONICS DIV.
4807 Bradford Drive
Huntsville, AL 35808
205/837-6500

KIMBALL INDUSTRIES, INC.
132 West Chestnut Avenue
Monrovia, CA 91016
818/357-6061.

L. LEE CONSULTANTS, Inc.
Box 9
Westport, CT 06881
203/366-7735

M-RAD — see Sec. 35.1.1.

NATIONAL TECHNICAL SYSTEMS
26525 Golden Valley Road
Saugus, CA 91350
805/251-6222.

TURN KEY SYSTEMS
136 South 8th Avenue
Industry, CA 91746
818/968-7743

YANKEE MAGCAST Co.
Box 2313
Enfield, CT 06082
203/749-8443

35.2 References

35.2.1 SVIC monographs. See Sec. 35.3.2.

Lyon, Richard H., *Random Noise and Vibration in Space Vehicles* (SVM-1), 1967.

Mustin, Gordon S., *Theory and Practice of Cushion Design* (SVM-2), 1968.

Enochson, Loren D., *Programming and Analysis for Digital Time Series Data* (SVM-3), 1968.

Loewy, Robert G. and Piarulli, Vincent J., *Dynamics of Rotating Shafts* (SVM-4), 1969.

Kelly, Ronald D., *Principles and Techniques of Shock Data Analysis* (SVM-5), 1969.

Sevin, Eugene and Pilkey, Walter D., *Optimum Shock and Vibration Isolation* (SVM-6), 1971.

Ruzicka, Jerome E. and Derby, Thomas R., *Influence of Damping in Vibration Isolation* (SVM-7), 1971.

Curtis, Allen J., Tinling, Nickolas G. and Abstein, Henry T., Jr., *Selection and Performance of Vibration Tests* (SVM-8), 1971.

Fackler, Warren C., *Equivalence Techniques For Vibration Testing* (SVM-9), 1972.

Pilkey, Barbara and Walter, *Shock and Vibration Computer Programs* (SVM-10), 1975.

Bouche, Raymond R., *Calibration of Shock and Vibration Measuring Transducers* (SVM-11), 1979.

35.2.2 Texts. Here are some helpful texts:

Bendat, Julius S. and Piersol, Allan G., *Random Data: Analysis and Measurement Procedures*, Wiley-Interscience, 1971 (in revision for 1985 publication). Also, *Engineering Applications of Correlation and Spectral Analysis*, (1980).

Broch, Jens Trampe, *Mechanical Vibration and Shock Measurements*, Bruel & Kjaer Measurement Systems (Denmark), 2nd edition, 1979. Ask B&K about other handbooks.

Crandall, Stephen H. and Mark, William D., *Random Vibration in Mechanical Systems*, Academic Press, 1963.

Crandall, Stephen H., editor, *Random Vibration*, M.I.T. Press, 1963.

Harris, Cyril E. and Crede, Charles E., editors, *Shock and Vibration Handbook* First Edition, in three volumes, McGraw-Hill, 1961. Probably out of print, but in some ways superior to

Harris, Cyril E. and Crede, Charles E., editors, *Shock and Vibration Handbook* Second Edition, in one volume, McGraw-Hill, 1976.

Keast, David N., *Measurements in Mechanical Dynamics*, McGraw-Hill, 1967.

Macduff, John N. and Curreri, John R., *Vibration Control*, McGraw Hill, 1958.

Morrison, Ralph, *Grounding and Shielding Techniques in Instrumentation*, John Wiley and Sons, 1967.

Morrow, Charles T., *Shock and Vibration Engineering* (Volume 1), John Wiley and Sons, 1963.

O'Connor, P.D.T., *Practical Reliability Engineering,* Heyden, 1981.

Salter, John P., *Steady-State Vibration*, Kenneth Mason (London), 1969.

Seippel, Robt. G., *Transducers, Sensors and Detectors*, Reston Publishing, 1983.

Steinberg, Dave S., *Vibration Analysis for Electronic Equipment*, John Wiley and Sons, 1973.

Thomson, William T., *Theory of Vibration with Applications*, Prentice-Hall, 1981.

Ask GenRad about handbooks dealing with vibration and sound.

35.2.3 Magazines. Besides the IES and SVIC publications mentioned above and below, you may wish to consider subscribing to

Evaluation Engineering, edited and published by A. Verner Nelson, 2504 N. Tamiaimi Trail, Nokomis, FL 33555.

Measurements and Control, edited and published by Milton H. Aronson, 2994 Liberty Avenue, Pittsburgh, PA 15216.

Sound and Vibration, edited and published by Jack Mowry, Acoustical Publications, Inc., 27101 E. Oviatt Road, Bay Village, OH 44140.

Test Engineering and Management, edited and published by Ray Mattingley, 61 Monmouth Road,Oakhurst, NJ 07755.

35.3 Technical organizations:

35.3.1 Institute of Environmental Sciences (IES). The IES is located at 940 East Northwest Hwy., Mt. Prospect, Illinois, 60056, 312/255-1561. The IES promotes understanding of the effects of environmental factors (both natural and induced) as they relate to the quality, producibility, reliability and safe operation of products, components and systems. National technical meetings are generally in April. The concurrent Technical Equipment Exposition is the largest USA "trade show" in the environmental field. IES has over 2,000 members. 23 local chapters meet frequently. Topics include stress screening of electronics production, design for random vibration, contamination control, environmental standards, space vehicle testing, reliability, shock and vibration, solar technology, technical writing, space simulation and earth resources.

35.3.2 The Shock and Vibration Information Center (SVIC) is located at Code 5844, U.S. Naval Research Laboratory in Washington, D.C. 20375, telephone 202/767-2220. SVIC is jointly sponsored by the Air Force, Army and Navy, also NASA. Firms contribute financially to help defray costs of SVIC's activities. A Shock and Vibration Symposium meets each October; papers appear in an annual *Shock and Vibration Bulletin*. The monthly *Shock and Vibration Digest* is edited by Dr. Ron Eshleman. You can write or telephone SVIC for technical assistance.

35.3.3 Other professional organizations somewhat concerned with shock and vibration:

The American Society of Mechanical Engineers   ASME
    345 E. 47th Street, New York, NY 10017

The American Inst. for Aeronautics & Astronautics   AIAA
    1290 Avenue of the Americas, New York, NY 10104

Society of Automotive Engineers   SAE
    21000 W. Ten Mile, Southfield, MI 48075

The American Society for Testing Materials   ASTM
    1916 Race Street, Philadelphia, PA 19103

The Society of Environmental Engineers   SEE
    Owles Hall, Buntingford, Herts., SG9 9PL ENGLAND

35.3.4 Schools. Training is offered by some equipment manufacturers. "Hands on" training using that manufacturer's equipment is highly valued. Some universities offer short, specialized courses during holiday periods, some taught by specialists from industry.

A very few specialized schools restrict their training to vibration and shock testing, measurement, analysis, calibration and stress screening.

The Vibration Institute, Inc.   Tustin Institute of Technology, Inc.
Dr. Ron Eshleman   Wayne Tustin
101 West 55th Street   22 East Los Olivos Street
Clarendon Hills, IL 60514   Santa Barbara, CA 93105
312/654-2254   805/682-7171

INDEX